Design and Installation of Marine Pipelines

Mikael W. Braestrup (Editor)

Jan Bohl Andersen

Lars Wahl Andersen

Mads Bryndum

Curt John Christensen

Niels-Jørgen Rishøj Nielsen

Blackwell
Science

© 2005 by Blackwell Science Ltd, a Blackwell Publishing company

Editorial offices:
Blackwell Science Ltd, 9600 Garsington Road, Oxford OX4 2DQ, UK
  Tel: +44 (0) 1865 776868
Blackwell Publishing Professional, 2121 State Avenue, Ames, Iowa 50014-8300, USA
  Tel: +1 515 292 0140
Blackwell Science Asia Pty Ltd, 550 Swanston Street, Carlton, Victoria 3053, Australia
  Tel: +61 (0)3 8359 1011

First published 2005

Library of Congress Cataloging-in-Publication Data
    Design and installation of marine pipelines / M.W. Braestrup . . . [et al.]. – 1st ed.
        p. cm.
    Includes bibliographical references and index.
    ISBN 0-632-05984-2 (hardback : alk. paper)
    1. Underwater pipelines.   I. Bræstrup, Mikael W.   II. Title.
TJ930.D378 2005
621.8'672'09162—dc22

                                                                        2004017651

ISBN-10: 0-632-05984-2
ISBN-13: 978-0632-05984-3

A catalogue record for this title is available from the British Library

For further information on Blackwell Publishing, visit our website:
www.thatconstructionsite.com

Contents

Preface

The construction of marine pipelines is a relatively new field of engineering, having developed in the course of the last five decades. Yet, with more than 100 000 km of subsea lines installed worldwide it must be regarded as a mature technology, although advances continue in extending the limits as regards pipe dimensions, pressures, flow regimes, products, installation methods, water depths and climatic environments. It is therefore surprising that there is a shortage of books that cover the entire process of marine pipeline design and installation, from project planning through to system operation.

Marine pipelines are generally designed, fabricated and installed in accordance with guidelines issued by various certifying agencies and regulatory bodies, as well as national and international codes. Ideally, however, a code should be a fairly slim volume, based upon functional performance criteria. The choice of design methods and construction procedures should be left to the engineers responsible, relying on a general consensus of good engineering practice, backed up by shared literature and education experience. However, this is precisely the material that is generally lacking in the marine pipeline field.

The aim of this publication is therefore to complement the existing codes and recommendations with an engineering book, serving as a guide to the profession with the objective of ensuring a reasonable standard of design and application. The book is primarily aimed at engineers who are fairly new to the field of marine pipelines, and want a comprehensive overview of the subject. The book should also provide background reading to students on specialised offshore courses, and to professionals in related fields. To experienced pipeline engineers it would constitute a reference work to be consulted for fact and figures.

The origins of the work go back to 1987, when it was felt that there was a need for a Danish national code for marine pipelines, and a code-drafting working party was established, eventually comprising most of the authors of the present book, chaired by the editor. The Danish marine pipeline code never materialised, but the working party identified the need explained above, and began collating supporting documentation, which eventually evolved into the present textbook. The result constitutes a set of comprehensive guidelines for the design and installation of marine pipeline systems, but the authors take no responsibility for the use of the recommendations given.

The authors acknowledge the valuable comments to the draft manuscript offered by Robert Inglis of J P Kenny. It should be stressed, however, that the responsibility for any errors or omissions rests with the authors alone. Finally, the authors would like to thank Dansk Olie og Naturgas A/S (DONG A/S) for permission to use photographs from the 1998–99 installation of the South Arne to Nybro 24-inch Offshore Gas Pipeline.

Mikael W. Braestrup
Jan Bohl Andersen
Lars Wahl Andersen
Mads B. Bryndum
Curt J. Christensen
Niels J. Rishøj Nielsen

About the authors

Editor

Mikael W. Braestrup, MSc, PhD
Born in 1945, Dr Braestrup obtained his PhD in structural engineering from the Technical University of Denmark in 1970. After spending two years as a volunteer in charge of low-cost road construction in Peru he was engaged in structural concrete research and teaching in Copenhagen, Denmark, and Cambridge, UK. He joined the consulting company RAMBOLL in 1979, and worked for five years on the design and construction of the offshore pipelines of the Danish Natural Gas Transmission System. Subsequently he has headed the RAMBOLL departments of Marine Pipelines and Underwater Technology, and of Knowledge and Development.

Since 1992 Dr Braestrup has been attached to the RAMBOLL Department of Bridges, but is also active in the field of marine pipelines. Recent assignments include coating design for the 24″ Danish North Sea gas trunkline and the 30″ BalticPipe between Denmark and Poland, feasibility studies for a 42″ gas transmission pipeline across the Baltic Sea from Russia to Germany, and secondment to the Statoil engineering team for the export pipelines of the Kristin HP/HT subsea development. Dr Braestrup is an active member of a number of international associations (*fib* (CEB-FIP), IABSE, ACI), has served on several Danish code committees, and is the author of a substantial number of papers and reports on civil engineering subjects, including marine pipelines.

Co-authors

Jan Bohl Andersen, MSc, started his professional career at the RAMBOLL pipeline department, and then spent some years as an independent consultant, mostly in Norway. Since 1993 he has been back with RAMBOLL, specialising in design, contracting, and project management of marine pipelines.

Lars Wahl Andersen, MSc, worked in the RAMBOLL pipeline department until 1992, when he transferred to the Department of Risk and Reliability. His expertise on safety and risk studies is widely applied to offshore projects, platforms as well as pipelines.

Mads Bryndum, MSc, is chief engineer in the Ports and Offshore Department at DHI Water and Environment. Mads Bryndum has a background in structural engineering and worked for several years at C G DORIS on the design of concrete gravity platforms. He joined DHI in 1980 and has specialised in hydraulic and hydrodynamic problems in relation to marine pipelines and risers.

Curt John Christensen, BSc, has been employed by FORCE Technology (formerly the Danish Welding and Corrosion Institute) since 1976 as a corrosion and metallurgy specialist. He started in the marine engineering sector and, from 1980, he has been deeply involved in oil and gas industry and pipeline industry related jobs. From 1985 to 2000 he was in charge of the full scale pipeline testing activities at FORCE Technology.

Niels-Jørgen Rishøj Nielsen, MSc, PhD, heads the Engineering and R&D Department of NKT Flexibles I/S. Following a PhD in structural optimisation of ship structures, he began his professional offshore career in 1984 at the Danish Maritime Institute. He then moved to Maersk Olie og Gas A/S, working in their pipeline department, and spent some years at the DNV Copenhagen office before joining NKT Flexibles in 1995.

Glossary and notation

Introduction

Like most other specialised fields the jargon of the offshore pipeline world is rife with acronyms and terms whose specific meaning is not obvious to the uninitiated. The sections below list the employed abbreviations and explain the frequently used symbols and special terms.

SI units are used throughout, with a few exceptions, such as bar to indicate pressures and inch (") to designate pipe sizes. However, all equations are dimensionally correct, unless specifically noted.

Symbols

Specific symbols are defined in the text, and a comprehensive, global listing would have little value as the meaning often depends on the context. However, the list below contains the most common notations as they are used in this book, with or without descriptive indices. A few symbols (m, α, β) are also used to designate sundry constants, in addition to the specific meanings given below.

A area, amplitude, elongation (material property)
D pipe diameter, grain size, diffusion coefficient (material property)
E modulus of elasticity (material property)
F force
f strength, skin friction factor (material properties), frequency of occurrence
g acceleration of gravity
H height
h water level
k thermal conductivity, roughness (material properties)
L length
l length
m mass

N	axial force
P	probability, permeability (material property)
p	pressure, load
R	reaction force, anodic resistance, steel strength (material property)
r	radius
T	time
t	pipe wall thickness, time
U	water particle velocity
v	velocity
W	weight
α	coefficient of thermal expansion (material property)
β	reliability index (or safety index)
Δ	deformation, increment or variation
ε	strain
ϕ	angle of friction (material property)
γ	safety factor
η	usage factor
μ	coefficient of friction (material property)
ν	Poisson's ratio, viscosity (material properties)
θ	temperature
ρ	density, specific ohmic resistivity (material properties)
σ	stress

Abbreviations

This list defines the acronyms used in this book. Sections giving further explanations are referenced in brackets, if relevant. Chemical elements, units or notations are not included.

3-RF	3-Roller Forming (7.2.4)
ADCP	Acoustic Doppler Current Profiler (2.5.3)
AGA	American Gas Association (4.3.3)
AISI	American Iron and Steel Institute (10.2.2)
ALARP	As Low As Reasonably Practicable (5.4.1)
ALS	Accidental Limit State or Accidental Load Situation (5.2)
API	American Petroleum Institute
ASD	Allowable Stress Design (6.1.1)
ASM	American Society of Metals (7.2.1)
ASME	American Society of Mechanical Engineers
ASTM	American Society for Testing and Materials
BOP	Blow Out Preventer (4.4.2)

BS	British Standard
CAPEX	Capital Cost (2.3.3)
CBV	Cost Benefit Value (5.4.1)
CEN	Centre Européen de Normalisation
CEV	Carbon Equivalent (7.2.5)
CIV	Corrected Inherent Viscosity
COE	C-bending, O-pressing, Expansion (7.2.4)
CP	Centre Point (2.6.2), Cathodic Protection
CPT	Cone Penetrometer Testing (2.4.4)
CRA	Corrosion Resistant Alloy (3.3.1)
CTOD	Crack Tip Opening Displacement
CSA	Canadian Standards Association
DEH	Direct Electrical Heating (6.6.5)
DFI	Design, Fabrication and Installation (9.6.1)
DFO	Documents For Operation (9.1)
DFT	Dry Film Thickness (7.3.4)
DIN	Deutscher Institut für Normung
DNV	Det Norske Veritas
DONG	Dansk Olie og Naturgas
DP	Dynamic Positioning (8.4.1)
DS	Dansk Standard
DWT	Dead Weight Tonnage
DWTT	Drop Weight Tear Test (7.2.5)
ECA	Engineering Criticality Assessment (8.3.3)
EEZ	Exclusive Economic Zone (2.2.3)
EFC	European Federation of Corrosion
EIA	Environmental Impact Assessment (2.2.3)
EN	Euro Norm
ENV	European Pre-Standard (draft EN)
ESD	Emergency Shut Down (valve) (4.2.3)
EU	European Union
FAR	Fatal Accident Rate (5.4.1)
FBE	Fusion Bonded Epoxy (7.3.3)
FEM	Finite Element Method
FLS	Fatigue Limit State (6.1.1)
FMEA	Failure Mode Effect Analysis
FPS	Floating Production System (10.1.1)
FPSO	Floating Production, Storage and Offloading (10.3.5)
FRP	Fibre Reinforced Polymer
GA	General Arrangement (drawing)
GIS	Geographical Information System (9.6.2)
GL	Germanischer Lloyd
GMAW	Gas Metal Arc Welding (8.3.2)

GPS	Global Positioning System
GPTC	Gas Piping Technology Committee (4.2.3)
HAZ	Heat Affected Zone
HAZID	Hazard Identification (5.3.2)
HAZOP	Hazard and Operability (study) (5.3.2)
HD-	High Density (e.g. HDPE, HDPU)
HDD	Horizontal Directional Drilling (8.5.5)
HFW	High Frequency Welding (7.2.4)
HIC	Hydrogen Induced Cracking (3.3.3)
HIP	Hot Isostatic Pressing (7.2.7)
HIPPS	High Integrity Pressure Protection System (4.2.3)
HP/HT	High Pressure/High Temperature
HSE	Health, Safety and Environmental Protection (9.2.2)
ID	Internal Diameter (pipe bore)
IEC	International Electrotechnical Commission (4.2.3)
IP	Intersection Point (2.6.2)
ISO	International Standardisation Organisation
ISO/DIS	Draft International (ISO) Standard
KP	Kilometre Post (2.6.2)
LAT	Lowest Astronomical Tide
LCI	Life Cycle Information (9.1)
LI	Liquidity Index (3.3.2)
LL	Liquid Limit (3.3.2)
LRFD	Load and Resistance Factor Design
MAIP	Maximum Allowable Incidental Pressure (4.2.3)
MAOP	Maximum Allowable Operating Pressure (4.2.3)
MBR	Minimum Bending Radius (10.3.3)
MD-	Medium Density (e.g. MDPE)
MEG	Methyl Ethyl Glycol/Mono Ethylene Glycol
MPI	Magnetic Particle Inspection
MSL	Mean Sea Level
MTO	Material Take-Off
NACE	National Association of Corrosion Engineers (USA)
NDE/NDT	Non-Destructive Examination/Non-Destructive Testing
NGL	Natural Gas Liquids
NKT	Nordisk Kabel og Tråd (10.1.2)
NORSOK	Norsk Sokkels konkuranseposisjon (competitive standing of the Norwegian offshore sector)
NPV	Net Present Value (2.3.3)
OD	Outer Diameter
OPEX	Operational Cost (2.3.3)
OREDA	Offshore Reliability Data (5.7)
OS	Offshore Standard

P&ID	Process and Instrumentation Diagram
PA	Polyamide
PD	Published Document (6.1.1)
PE	Polyethylene
PI	Plasticity Index (3.3.2)
PL	Plastic Limit (3.3.2)
PP	Polypropylene
PRCI	Pipeline Research Council International, Inc. (4.3.3)
prEN	European Pre-Standard
PSD	Process Shut Down (valve) (4.2.3)
PSV	Pressure Safety (relief) Valve (4.2.3)
PU	Polyurethane
PVC	Poly Vinyl Chloride
PVDF	Polyvinylidene Fluoride
QA/QC	Quality Assurance/Quality Control
QRA	Quantified Risk Analysis (5.3.2)
REM	Rare Earth Materials (3.3.2)
RFO	Ready For Operation (8.8.1)
ROV	Remotely Operated Vehicle
RP	Recommended Practice
RTS	ROV-operated Tie-in System (8.6.4)
S	Seamless (pipe) (7.2.2)
SAR	Synthetic Aperture Radar (2.5.3)
SAW	Submerged-Arc Welding
SAWH	Submerged-Arc Welding Helical
SAWL	Submerged-Arc Welding Longitudinal
SCADA	Supervision, Control and Data Acquisition (4.2.3)
SCF	Stress Concentration Factor
SI	Système Internationale/International System of Units of Measurements
SLOR	Single Line Offset Riser (10.3.5)
SLS	Serviceability Limit State (6.1.2)
SMAW	Shielded Metal Arc Welding (8.3.2)
SMSS	Super Martensitic Stainless Steel
SML	Seamless (pipe) (7.2.2)
SMTS	Specified Minimum Tensile Strength
SMYS	Specified Minimum Yield Stress
SOHIC	Stress Oriented Hydrogen Induced Cracking (3.3.3)
SRB	Sulphate Reducing Bacteria (8.8.2)
SSC	Sulphide Stress Cracking (3.3.3)
SSIV	Subsea Safety Isolation Valve
SWC	Stepwise Cracking (3.3.3)
SZC	Soft Zone Cracking (3.3.3)

TEG	Tri-Ethylene Glycol
TLP	Tension Leg Platform
TFL	Through Flowline
TMCP	Thermo-Mechanical Controlled Processing (3.3.1)
TP	Tangent Point (2.6.2)
TÜV	Technische Überwachungs Verein
ULS	Ultimate Limit State (6.1.3)
UNECE	United Nations Economic Commission for Europe (2.2.3)
UNS	Unified Numbering System
UOE	U-shaping, O-pressing, Expansion (7.2.4)
UTC	Unit Transportation Cost (2.3.3)
UV	Ultraviolet (radiation)
VIV	Vortex Induced Vibrations (6.4.6)
WBC	Wet Buckle Contingency (8.4.6)
WBS	Work Breakdown Structure
WC	Water Content (3.3.2)
WT	Wall Thickness (steel)
XLPE	Cross-linked Polyethylene (10.4.3)

Terms

The list explains the most common specialised terms, in the sense in which they are employed in this book, as well as a few other terms in common usage.

Acid wash
Chemical treatment of steel substrate by means of diluted mineral acid (usually phosphoric acid) in combination with detergency systems.

Actuator
See *Ball valve*.

Added mass
Additional mass is assigned to a body to reflect the fact that the force required to accelerate a submerged body is larger than the force required to accelerate the same body in vacuum. The reason for this is that part of the surrounding water is accelerated together with the body. The added mass is typically described by a coefficient and the mass of the volume of water displaced by the body.

Alignment sheet
Drawing showing a section of pipeline (in plan and profile), incorporating seabed features as well as physical pipeline properties and installation parameters.

Anode

Electrode from which electric current flows to an electrolyte (water, soil). On the surface an oxidation process takes place, e.g. metal to metal ions or hydroxyl ions to oxygen and water.

Anode assembly

Also referred to as anode bank. A cluster of anodes delivering current, e.g. to provide cathodic protection. Anodes placed in (onshore) soil are normally surrounded by low resistance material (e.g. coke), and are referred to as anode beds.

Anode pad

See *Doubler plate.*

Asphalt

See *Bitumen.*

Asphaltene

Non-volatile, high molecular weight fraction of petroleum that is insoluble in light alkanes such as n-pentane and n-heptane.

Atterberg Limits

Limiting water contents between which soil behaviour is characterised as a plastic solid (as opposed to a semisolid or a viscous liquid).

Austenite

See *Steel microstructure.*

Backfilling

Covering of trenched pipeline, which may be natural (by sedimentation) or artificial (by rock dumping or by mechanically returning the seabed material removed during trenching).

Bainite

See *Steel microstructure.*

Ball valve

Valve where the valve body is built around a ball provided with through hole, the size of the pipe bore. The valve is closed by rotation of the ball manually or by means of a hydraulic actuator, which may be diver, ROV or remotely operated.

Barrier coating

Blockage against oxygen and other gases, as well as water and dissolved salts.

Bend

Curved piece of pipe, for offshore use either hot formed from induction bent linepipe joints (motherpipe) or forged items. Bends with small bending radius ($1.5 \times$ ID or less) are also referred to as elbows, and will normally be forged. To facilitate the welding into the pipeline, bends are normally provide with short, straight sections (tangent lengths).

Bellmouth

Part of a guide tube, formed in the shape of a bellmouth, to control the bending curvature of the flexible pipe in dynamic or static applications.

Bending restrictor

A mechanical device that functions as a mechanical stop to limit the bending curvature of a flexible pipe in static applications.

Bending stiffener

A cone-shaped ancillary component that supports the flexible pipe and thus prevents over-bending of the pipe in dynamic or static applications.

Benthic

Generic term for plant and animal life living at the seabed.

Bitumen

Coating material derived from distillation of hydrocarbons or extracted from natural deposits (asphalt).

Bleeding

See *Venting*.

Bonded flexible pipe

A flexible pipe where the steel reinforcement is integrated and bonded to a vulcanised, elastomeric material.

Bottom tow

Installation method whereby pipe strings are pulled in position on the seabed (see *Towing*).

Buckle

Deformation of pipeline as a result of local actions or stability failure of the pipe section due to external pressure, possibly in combination with bending. The buckling may lead to water entering the pipeline (wet buckle) or not (dry buckle).

Buckle arrestor
Section of thick-walled linepipe introduced at regular intervals to prevent propagation buckling of a pipeline.

Caisson
Sleeve pipe housing one or several pipelines, cables, etc.

Cathode
Electrode into which electric current flows from an electrolyte (water, soil). On the surface a reduction process takes place, e.g. water to hydrogen and hydroxyl ions.

Cathodic disbonding
Loss of bond between barrier coating and steel substrate due to the formation of hydroxyl ions in connection with cathodic protection.

Cathodic protection
Electrochemical method of corrosion protection of metal structures, achieved by forcing an electric current from an anode through the surrounding electrolyte into the metal, which becomes a cathode.

Chainage
Distance measured along the pipeline by accumulating the lengths of the installed pipe joints. Owing to local deviations and seabed irregularities this will be somewhat higher than the kilometre post (KP) reckoning along the theoretical pipeline route. Thus if KP numbers are used to designate chainage it should be highlighted.

Charpy V test
Test method to determine impact fracture properties of steels.

Check valve
Non-return valve, preventing counterflow. Swing check valves (also known as flapper or clapper valves) are built around a flapper disc, attached at the valve house top, which can be rotated to close the valve. During operation the disc is suspended by the fluid flow, and if there is a change in flow direction the flapper will swing closed, providing a seal that will prevent fluid loss.

Chromate conversion coating
Chemical treatment of steel substrate by wetting the surface with diluted chromate acid in combination with other chemicals.

Cladding
See *Sheeting*.

Clapper valve
See *Check valve*.

Crack arrestor
Device to be incorporated in a pipeline to prevent long running crack.

Coal tar
Coating material manufactured by distillation (pyrolysis) of rock coal.

Coating yard
Onshore facility for the application of pipe coatings and sacrificial anodes to pipe joints.

Cobbles
See *Rock dumping*.

Cold springing
Introduction of residual moment in a connection by elastic deformations during tie-in.

Collapse
Deformation of a pipeline due to a distributed load, particularly external pressure.

Components
Pressure sustaining parts of a pipeline system which are not linepipe (e.g. fittings, valves, isolation couplings, pig launchers/receivers).

Concrete coating
Pipe coating of reinforced concrete, applied to increase the pipeline weight and/or protect the steel pipe and its anti-corrosion coating against mechanical damage.

Condensate
Liquid hydrocarbon, separated from natural gas by reduction of pressure and temperature. Also referred to as NGL.

Contractor
Executing party in a contractual relationship, including sub-contractor or supplier.

Counteract
Lateral support on the seabed, guiding the pipeline during pipelaying in horizontal curves, also called a turnpoint.

Corrosion allowance
Increase of the wall thickness corresponding to the expected corrosion loss, with the objective of ensuring the required wall thickness during the service life.

Crossing

Intersection of pipeline with a previously installed (operational or abandoned) pipeline or cable.

Crushing

Radial compression, which may cause a sudden collapse or significant ovalisation of the pipe cross-section.

Cutting

Trenching of pipeline by means of a mechanical digging machine riding on the pipe, towed or self-propelled.

Davit lift

Above-water tie-in operation, involving lifting of the two pipe string ends by barge-mounted cranes (davits).

Dead man anchor

High holding anchor with corresponding chain, placed on the seabed to provide tension for pipelay initiation.

Dent

Local deformation of the linepipe wall, resulting in a reduction of the pipe bore.

Design pressure

Maximum internal pressure occurring in the pipeline during normal operation, referred to a specific reference height.

Directional drilling

Installation method whereby the pre-fabricated pipe string is pulled through a hole drilled through the soil.

Dogleg

See *Expansion offset*.

Double jointing

Welding together of two pipe joints before they are incorporated into the pipe string.

Doubler plate

Steel plate welded on to the pipe joint under factory conditions, to provide a location for structural attachment, in particular anode pads for sacrificial anode cable connection, by stick welding, pin brazing or thermite welding.

Drop weight tear test
Test method to determine fracture properties of heavy sections of steel.

Dry buckle
See *Buckle*.

Duplex steel
Steel with mixed microstructure. Stainless duplex steel is a chromium alloyed mixture of austenite and ferrite. Normal grade is 22 Cr, the higher grade 25 Cr with higher strength and superior corrosion resistance is also referred to as super duplex.

Dye stick
Solid dye units designed to dissolve after flooding, and serve in case leak detection is required. Dye sticks are easy to place in critical units such as tie-in or valve spools. The dye can be coloured or clear but fluorescent.

Eddie current testing
Non-destructive test method based on electromagnetic principles.

Elbow
See *Bend*.

Elbolet
See *Olet*.

Enamel
Hot applied pipe coating consisting of bitumen or coal tar, reinforced with layers of fibreglass wrap.

Epoxy paint
Two-component paint consisting of epoxy resin and solvent.

Expansion buckling
Lateral deformation, possibly of large displacement, of a pipeline, caused by prevented longitudinal expansion to relieve compressive forces due to increases in pressure and/or temperature.

Expansion offset
Pipe spool (in the shape of an L, Z or U) that is inserted between a pipeline and a fixed structure (e.g. a platform riser or a wellhead) to absorb longitudinal deformations due to changes in temperature or pressure of the pipeline medium. Also referred to as an expansion loop or dogleg.

Export pipeline
Pipeline transporting a treated product (gas, condensate, oil) from an offshore facility. It may be an interfield pipeline or a transmission pipeline.

Fall pipe
Vertical steel pipe suspended from a surface vessel, through which rock or gravel is dumped on the seabed. See *Rock dumping*.

Ferrite
See *Steel microstructure*.

Field joint
Connection of two pipe joints carried out on site (offshore or at an onshore site or construction yard). It includes the girth weld and the field joint coating, bridging the gap between the factory-applied anti-corrosion coating on the outside, and possibly also the inside, of the pipe. Additional field joint infill may fill the gap between insulation or concrete coating on the adjoining pipe joints.

Firing line
Work area on a laybarge where the pipe joints are welded onto the pipe string.

Fish
Streamlined platform or container for survey, and other, instruments; towed by a surface vessel along a specified route, used for data gathering.

Fittings
Pipeline components that do not have operational functions (e.g. flanges, tees, wyes, olets).

Flapper valve
See *Check valve*.

Flange connection
See *Mechanical connection*.

Flexible pipe
Factory produced pipe string characterised by a layered configuration, resulting in a bending stiffness which is orders of magnitude smaller than for steel pipe of similar dimension.

Flooding
Filling the pipeline with water, to perform hydrotesting or to facilitate tie-in.

Flotation
Upwards directed, vertical instability of a buried pipeline.

Flow assurance
The prevention of production losses by ensuring unrestricted flow path in the pipe transportation system during its service life.

Flowline
Pipeline transporting untreated well fluids.

Flow regime
Nature of the transported medium. Normal designations are: single phase (liquid or gaseous), two-phase (e.g. oil and gas) and multi-phase (e.g. oil, gas, water and particles).

Fracture toughness
Measure of the ability to resist crack propagation under sustained load. See also *Impact toughness*.

Free span
Section of pipeline unsupported by the seabed.

Girth weld
Circumferential weld between pipe joints. See also *Field joint*.

Gouge
Local linepipe imperfection, affecting the pipe wall only, without resulting in a reduction of the pipe bore.

Gravel
See *Rock dumping*.

Heat affected zone
Zones adjacent to welds where the heat input from welding changes the base metal microstructure.

Heat shrink sleeve
Field joint anti-corrosion coating consisting of cross-linked polyolefin, which shrinks upon the application of heat.

Hot spot stress
Imaginary reference stress for welded joints. Hot spot stress is established by extrapolation of stresses outside the weld notch zone into the singularity at the

weld root or toe. Hot spot stress includes geometric stress concentration, but not the high local stress generated by the welding process.

Hot-tapping
Connection of a branch line to an existing pipeline without emptying the latter, as an alternative to using a pre-installed tee (or wye).

Hub
Short piece of pipe with a raised flange to engage with a remotely operated tool (e.g. for tie-in or hot-tapping).

Hydrates
Ice-like combinations of water and natural gas, formed for certain temperature and pressure conditions, depending on the gas composition.

Hydrogen embrittlement
Damage to steel material incurred through the uptake of atomic hydrogen, e.g. in connection with welding, galvanising, cathodic protection or corrosion.

Hydrotesting
Short for hydrostatic testing, whereby the strength and tightness of a pipeline section is documented by flooding with water and pressurising.

Hyperbaric welding
Welding performed subsea in a pressurised habitat. See also *Saturation diving*.

Impact toughness
Measure of the ability to resist crack initiation and propagation under high impact loads. See also *Fracture toughness*.

Impingement
Application of concrete coating whereby a no-slump concrete mix is thrown at the rotating pipe joint.

Impressed current
Method of cathodic protection where the driving current is delivered from an external power source.

Incidental pressure
Maximum internal pressure that can occur in the pipeline during operation, referred to the same reference height as the design pressure.

Infill
Field joint coating material filling the gap between the coatings on the adjacent pipe joints.

Inhibitor
Substance added to a transported medium to reduce corrosion.

Initiation head
See *Laydown head*.

Intelligent pig
Pig equipped with sensors and recording devices used for internal inspection of pipelines.

Interfield pipeline
Pipeline between offshore structures or installations.

Isolation coupling
Pressure sustaining device providing electrical isolation of two pipeline sections.

Jetting
See *Water-jetting*.

J-laying
Pipelaying where the pipe string leaves the laybarge in a vertical or nearly vertical position.

J-tube
A pre-installed vertical, or near vertical, protection tube having a straight section with a bend at the lower end, through which a riser may be pulled. If the bend is omitted the term I-tube is used.

Jumper
Short flexible pipe used subsea and topside in static or dynamic applications.

Landfall
See *Shore approach*.

Landline
Onshore pipeline.

Latrolet
See *Olet*.

Laybarge
Vessel (whether self-propelled or not) used for pipelaying.

Laydown head

Piece of pipe with a wire attachment, welded on to the end of a pipe string to facilitate later retrieval. The laydown head at the start of pipelaying is called an initiation head, and may be provided with valves and pigs for dewatering of the line. To provide tension for pipelaying the initiation head is attached to a dead man anchor or to start piles. For a towed or pulled pipe string the term pull head is used.

Linepipe

Steel material for welded pipelines.

Lining

See *Sheeting*.

Liquefaction

Soil state where the effective stresses between the grains vanish, causing the soil to behave like a fluid.

Location Class

DNV classification of pipelines based on failure consequences due to location. Location Class 2 are areas (near platforms or landfalls) with frequent human activity, Location Class 1 is everywhere else. Also referred to as Zone 2, respectively Zone 1.

Magnetic particle inspection

Non-destructive test method based on magnetic fields around surface discontinuities.

Marine pipeline

Pipeline crossing a body of water (normally salt, but also fresh).

Martensite

See *Steel microstructure*.

Maximum wave height

Maximum height of waves in a sea state, measured as the vertical distances between succeeding wave troughs and wave crests.

Mechanical connection

Means of connecting pipes (tie-in) without welding, normally made by the bolting together of flanges welded to the two pipe strings. May involve proprietary coupling devices and tie-in tools.

Medium

Substance (gas, liquid, slurry) being transported through a pipeline. See *Flow regime*.

Meteo-marine
Meteorological and hydrographical (data or investigations). Also referred to as metocean.

Mill test pressure
Internal test pressure applied to linepipe joints during manufacture.

Miner's rule
See *Palmgren–Miner's rule*

Moonpool
Access to the sea (e.g. for divers) located inside a vessel.

Motherpipe
See *Bend*.

Non-return valve
See *Check valve*.

Normalising
Heat treatment of steels aimed at improving the final microstructure and increasing the mechanical properties.

Olet
Collar-shaped fitting welded on to the pipeline to reinforce a small bore branch connection. The connection may be welded (weldolet) or threaded (thredolet). Normally the branch is perpendicular to a straight pipe run, but it may be oblique (latrolet) or placed on a bend (elbolet).

Operating pressure
Internal pressure at which the pipeline is normally operated.

Ovalisation
Difference from a circle of the pipe cross-sectional geometry, usually arising from the fabrication process or as a result of bending. The ovality may be measured in percentage terms or as the difference between maximum and minimum diameters. Also referred to as out-of-roundness.

Overbend
Curved section of the pipe string supported by the laybarge stinger during S-laying.

Owner
Receiving party in a contractual relationship, pipeline operator.

Palmgren–Miner's rule
Method for predicting fatigue life under any type of variable amplitude loading. The method is based on a linear accumulation of the damage induced by each stress cycle.

Pearlite
See *Steel microstructure*.

Pig launcher/receiver
Facilities connected to a pipeline (separated by valves) for the dispatch and collection of pigs. Also referred to as pig traps.

Pig train
Series of different purpose pigs, separated by slugs of liquid.

Pigging
Passage of a pipeline by spherical or cylindrical devices (pigs), propelled by the transported medium. Pigs are used for cleaning, gauging, internal inspection (see *Intelligent pig*), or the separation of different transported media. Pigs can be uni-directional or bi-directional, the latter implying that the pig can be propelled from either end. Similarly, bi-directional pigging means pigging of the pipeline in both directions.

Piggy-backing
Installation method whereby a (smaller) pipeline is attached to another (larger) pipeline by strapping or clamping during pipelaying.

Pin brazing
Electrical cable connection made by brass soldering of a pin through a hole in the cable shoe.

Pipe jacking
Tunnelling method for installing underground pipelines with a minimum of surface disruption. The pipes are assembled at the foot of a shaft and jacked through the ground while the soil is removed from inside the encasement.

Pipe joint
Unit of linepipe manufactured at a pipe mill. For offshore application the length is normally 12.2 m (40′).

Pipe Mill
Production facility for pipe making.

Pipe string
Section of a pipeline welded together from pipe joints or manufactured as a flexible pipe. Also referred to as a stalk.

Pipe-in-pipe
Thermal insulation concept where insulation material is introduced in the annulus between the product pipe and a rigid sleeve pipe.

Pipelaying
Installation method whereby the pipe string is welded together on a laybarge as it is installed on the seabed.

Pipeline
Tubular conduit made from linepipe or flexible pipe transporting a medium, the driving force being a pressure differential between inlet and outlet.

Pipeline bundle
Several pipelines (and possibly cables) enclosed in a common caisson or sleeve pipe.

Pipeline system
Complete pipeline from inlet to outlet, including any risers, expansion offsets, valve assemblies, isolation couplings, spur lines, shore approaches, pig launchers/ receivers, etc.

Pitting
Localised corrosion attack.

Ploughing
Trenching of pipeline below the seabed by means of a plough riding on the pipe, normally towed. The plough may be used for backfilling as well, either simultaneously or consecutively.

Pour-point depressant
Additive for lowering the temperature at which the oil will pour or flow when being chilled.

Pre-commissioning
Activities (e.g. cleaning, de-watering, drying) which are required before a pipeline can be taken into operation (commissioned). Also referred to as RFO.

Pressure control system
System for controlling the internal pressure in the pipeline, comprising the pressure regulating system, the pressure safety system, and associated monitoring and alarm systems.

Pressure regulating system
System for ensuring that a set internal pressure in the pipeline is maintained.

Pressure safety system

System, independent of the pressure regulating system, for ensuring that the internal pressure in the pipeline does not exceed the allowable incidental pressure.

Pressure surge

Increase in pipeline pressure following a flow incident. A sudden valve closure or a decreasing pump outlet will result in a pressure surge.

Propagating buckle

Collapse of a large section of a pipeline (propagation buckling), due to external pressure following local buckling (see *Buckle*).

Propagating fracture

Fracture that runs apace for a significant length.

Pull head

See *Laydown head*.

Pup piece

Short spool piece welded onto component (e.g. valve or tee piece) to provide transition to the topical linepipe (diameter, wall thickness and steel quality).

Reeling

Installation method whereby the pre-fabricated pipe string, spooled on to a drum or reel, is paid out by a reel barge.

Ribbon anode

Long, thin sacrificial anode placed adjacent to the pipeline (in the soil or inside a caisson) to provide cathodic protection.

Ripple factor

Intensity of small fluctuating loads superimposed on (high) nominal (mechanical or current) loads.

Riser

Pipe string connecting an above-water offshore facility with a pipeline on the seabed. The tie-in to the pipeline (or expansion offset) may be direct or through an intermediate structure (riser base) on the seabed.

Riser guard

Steel structures, welded or bolted to the platform at the water line to protect risers against impact from vessels or floating debris.

Rock dumping
Placement of rock material on the seabed, e.g. as scour protection, support of a pipeline at a free span or crossing, or covering of a pipeline (trenched or untrenched) for protection or control of expansion or upheaval buckling. Rocks of sizes between 60 mm and 200 mm are called cobbles, whereas lower dimensions are referred to as gravel. The gravel fraction is further divided into coarse (20–60 mm), medium (6–20 mm) and fine (2–6 mm).

Sacrificial anode
An anode which connected to a structure can offer cathodic protection while it is consumed.

Safety Class
Classification of pipelines based upon the risk of human injury, environmental damage and economic loss. Pipelines are classified according to category of transported medium, location class, and duration of exposure (temporary or operational phase).

Sagbend
Curved section of the pipe string supported by the seabed during S-laying.

Saturation diving
Diving under (nearly) constant pressure, the diver alternating between working subsea, normally in a pressurised habitat, and living in a pressure chamber environment, obviating the need for depressurisation at the completion of each dive.

Scale formation
Forming of reaction products or settling of foreign matter on to surfaces.

Self-burial
Naturally occurring penetration into the seabed due to wave and current action.

Service line
Pipeline (normally small diameter) transporting an auxiliary medium (e.g. methanol or lift gas).

Sheeting
Sandwich combination of a carbon steel plate with a plate of corrosion resistant alloy. The process is referred to as cladding if there is metallic bond between the two layers. A CRA lining (typically inside a pipe) may or may not be metallically bonded to the carbon steel.

Shore approach
Connection between a marine pipeline and a landline, also referred to as a landfall.

Significant wave height

Statistical parameter characterising a sea state. The significant wave height was originally defined as the average wave height of the highest one third of the waves in a wave record. The significant wave height, H_s, is today in most cases calculated using the statistical parameters of the spectrum characterising the sea state ($H_s = 4\sqrt{m_0}$, where m_0 is the zero moment of the wave spectrum).

S-laying

Pipelaying where the pipe string leaves the laybarge in a horizontal or nearly horizontal position (as opposed to J-laying).

Sleeve pipe

See *Caisson*.

Slug catcher

Facility for separating and receiving liquid dropout in gas pipelines.

Slug flow

Intermittent gas and liquid flow regime in a pipeline or riser during operation.

Slurry

Transported medium consisting of solid particles suspended in a liquid.

Snaking

Horizontal, or nearly horizontal, expansion buckling, the pipeline sliding on the seabed.

Sour service

Operation of pipes or equipment in wet environments containing hydrogen sulphide, giving rise to sour corrosion.

Spool piece

Segment of pipe (including straight sections and/or bends) fitted between a pipe string and a fixed structure or component (e.g. riser, wellhead, valve assembly, pig launcher/receiver). See also *Pup piece*.

Spur line

Branch pipeline connected to the main pipeline by a tee or a wye.

Stalking

Installation of a riser already connected to a pipeline on to an offshore platform. See also *Pipe string*.

Start piles
Temporary structure on the seabed, installed by a subsea pile hammer or by the use of suction piles, to provide tension for pipelay initiation.

Steel microstructure
Grain structure of steel resulting from the manufacturing process. Typical microstructure designations are: austenite, martensite, ferrite, bainite and pearlite.

Stinger
Structure extending from the laybarge stern supporting the pipe string during S-laying.

Subsea
Generic term describing equipment or processes on the seabed.

Subsea completion
Hydrocarbon production unit (wellhead) located subsea.

Submerged weight
Weight in water, i.e. weight in air reduced by the buoyancy of the seawater.

Suction pile
Anchorage on the seabed provided by a cylinder, closed at the top and open at the bottom, which is evacuated to create an under-pressure.

Swan neck
S-shaped double bend, typically forming the vertical transition between a pipeline on the seabed and a raised component or structure.

Sweet service
Operation of pipes or equipment in wet environments containing carbon dioxide, giving rise to sweet corrosion.

Tangent length
See *Bend*.

Tee piece
Fitting for connection of a spur line perpendicularly to the main pipeline, usually permitting pigging of the main line, but not the spur line. Normally the complete tee will be provided with elbow, valve, flange, etc. Alternatively, the tee piece may be provided with a hub for connection of hot-tapping tool.

Tensioner
Device on a laybarge gripping the pipe string (by caterpillar tracks or similar) to keep it under tension during pipelaying.

Thredolet
See *Olet*.

Thermite welding
Electrical cable connection made by ignition of exothermic metal powder in a mould enclosing the cable end.

Tie-in
On-site connection of pipe strings to each other, to risers, to subsea completions, or to adjoining landlines.

Topside
Generic term describing equipment or processes above water on offshore facilities.

Towing
Installation method whereby pre-fabricated pipe strings or bundles are towed into position on the seabed.

Transition temperature
A specific temperature where the steel impact properties change from ductile to brittle behaviour.

Transmission pipeline
Pipeline (normally large diameter) transporting a treated medium (e.g. crude oil or sales gas), usually over long distances. Also referred to as a trunkline.

Trenching
Positioning of a pipeline below seabed level, with or without backfilling.

Trunkline
See *Transmission pipeline*.

Turnpoint
See *Counteract*.

Twin anodes
Two sacrificial anodes mounted adjacent to one another on the pipeline and functioning as one anode.

Umbilical
A bundle of helically or sinusoidally wound, small diameter chemical, hydraulic, and electrical conductors for power and control systems.

Unbonded flexible pipe
A flexible pipe consisting of separate polymeric and metallic layers, where the layers are allowed to move relative to each other.

Upheaval buckling
Vertical expansion buckling of a pipeline, normally through the soil cover.

Usage factor
Ratio between actual and allowable value of a given parameter, normally stress. Also referred to as utilisation ratio.

Valve assembly
One or more in-line valves (e.g. ball valve, check valve, SSIV), including any by-pass lines and supporting/protective structures.

Venting
Intentional release of pipeline content to the environment, usually through a valve. In the case of liquid media the term bleeding is used.

Water-jetting
Trenching of pipeline by means of pressurised water, delivered from a towed trench machine riding on the pipe. On a small scale compressed air (air-jetting) may also be used.

Waterstop
Inflatable bladder deployed in the flooded pipeline to seal the end during tie-in operations in a dry habitat. It may also seal a dry pipeline against flooding when the laydown head is removed (e.g. for tie-in).

Weight coating
See *Concrete coating*.

Weldolet
See *Olet*.

Weld-overlay
Build-up of surface layer by a welding process.

Wellhead
See *Subsea completion*.

Wet buckle
See *Buckle*.

Wet welding
Welding under water (or in wet environment) without the use of a dry habitat. See also *Hyperbaric welding*.

Wye piece
Fitting for connection of a spur line obliquely to the main pipeline, normally permitting uni-directional pigging of the main line, or both lines if the bores are not (too) different.

Zone 1, Zone 2
See *Location class*.

Chapter 1
Introduction

1.1 Introduction

Whereas the installation of pipelines for the transportation of liquids over land may be traced back to antiquity, the establishment of marine pipelines is a more recent development of the latter part of the twentieth century. The fuel line installed across the English Channel in 1944 to supply the allied troops during the Normandy landings is often cited as the first example. In fact, before the war small diameter oil export lines had already been installed in shallow waters off the US Gulf coast, and possibly also in Caddo Lake (Louisiana), off California and in the Caspian Sea, where offshore hydrocarbon exploration began.

The first oil-producing well 'out of sight of land' (in the Mexican Gulf) was drilled in 1947, the first pipelay barge commissioned in 1952, and the first pipeline laid on the seabed in 1954. Separate tallying of offshore pipelines did not start until 1968, but during the following three decades it is estimated that close to 90 000 km of marine pipelines were installed for the transportation of hydrocarbons, with approximately 5000 km being added each year. The majority of the pipeline systems are located in the heavily developed regions of the Arabian Gulf, the Gulf of Mexico and the North Sea.

Marine pipelines can be divided into different types. Relatively short and small diameter pipelines that transport crude hydrocarbons from a wellhead to a production or separation facility are designated flowlines, whereas the term export pipelines is applied to the lines that transport the produced hydrocarbons. Often small diameter service lines will transport auxiliary media (e.g. corrosion inhibitors, lift gas or injection water) in the other direction, whereas umbilicals provide power and signalling. Interfield pipelines is the generic term for lines used to transfer oil, gas or water between offshore installations within a limited area. Transmission pipelines or trunklines transport large quantities of oil or gas from an offshore complex to shore or between two landmasses.

A pipeline system is defined as a pipeline section extending from an inlet point, typically an offshore platform or an onshore compressor station, to an outlet point, typically another offshore platform or an onshore receiver station,

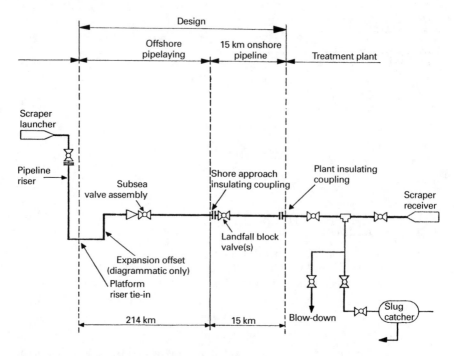

Figure 1.1   Schematic of a typical pipeline system

and the interfaces of the pipeline design normally follow this definition. Exceptional cases are spur lines connected to a main trunkline by tees or wyes, and pipelines terminating at subsea completions. Normally, facilities to launch vehicles (pigs) for cleaning, batching, and internal inspection of the pipeline are provided at the inlet point, whereas the outlet point has the facilities to receive these tools. Exceptions may include small diameter flowlines or service lines, or pipelines transporting non-corrosive media. Figure 1.1 presents a schematic outline of a typical pipeline system, in this case a transmission pipeline from an offshore platform to an onshore treatment plant.

The marine pipeline system design and installation covers the marine pipeline proper, platform risers, tie-in and spool connection parts, hydrostatic testing, possible subsea valve or branch assemblies, the corresponding protection works, as well as the activities conducted in association with start-up of production. A number of areas may be identified for special emphasis, such as crossings of other services (cables, pipelines), particularly hazardous zones of intense fishing or marine traffic, particularly sensitive environmental areas, etc.

As mentioned above, the construction of marine pipelines took off in the 1970s, and literature on the technology started to appear in the next decade, a prominent example being *Offshore Pipeline Design, Analysis, and Methods* (Mouselli 1981). A few years later *Advances in Offshore Oil and Gas Pipeline Technology* (de la Mare 1985) followed – a collection of contributions on various aspects of the subject. During the subsequent decades, however, the development of marine pipeline technology has mainly been reported at conferences, such as the annual Offshore Technology Conference (OTC), the biennial International Conference on Offshore Mechanics and Arctic Engineering (OMAE), the annual Offshore Pipeline Technology (OPT) European Seminar, and the biennial Pipeline Protection Conference. Other sources of information are lecture courses but, like conference proceedings, the documentation is not readily available to the public. Recent exceptions are *Pipelines and Risers* (Yong Bai 2001), which gives a comprehensive overview of design methods based upon two decades of research and teaching experience, and *Subsea Pipeline Engineering* (Palmer and King 2004). The latter scholarly monograph covers much the same ground as this book, but has less emphasis on practical construction issues.

This book addresses system layout determination, wall thickness design, route selection, fabrication, installation and tie-in methods, and pre-commissioning methodologies. The main emphasis is on structural design and construction, i.e. the assurance and verification that a safe and reliable pipeline system is established, once the basic parameters have been selected. System design (flow and capacity calculations) and operational issues are treated, but not to the same level of detail.

The text is descriptive in the sense that different methods and procedures are reviewed, and normative in the sense that specific requirements are stated and numerical coefficients proposed. The term marine implies that the pipelines are operating under water, including straits, estuaries and large rivers. The scope includes deep water as well as shallow water installations. The main emphasis is on transportation of hydrocarbons (oil, gas or multi-phase), but also media such as water, slurry and carbon dioxide are covered. However, issues specifically related to the transportation of potable water are not included. The scope is limited to pressure containment steel pipelines, including flexible lines; thus although plastic or cementitious materials may be used for marine crossings or coastal outlets, such pipelines are not covered, except by extension. The same holds for umbilicals and cables, where, for example, issues related to subsea couplings are not addressed.

The text is divided into eleven chapters. Following this introductory Chapter 1, the bases for design are reviewed in Chapter 2, including the overall requirements to function and environmental impact, the flow calculations, and the geotechnical and environmental site investigations, leading to the final route selection. Chapter 3 introduces topical materials: on the one hand the seabed soil (natural or engineered), on the other hand the linepipe steel, as well as sacrificial anodes and materials for coating, fittings and other components, etc.

Chapter 4 specifies the loads, which are divided into functional loads (internal pressure and temperature), environmental loads (principally waves and current), accidental loads (natural forces and third party interference), and loads incurred during installation. Risk and safety are treated in Chapter 5, the adopted safety format being limit state design combined with risk assessment and analysis, and operational risk management is discussed.

Design aspects are described in Chapter 6, which starts by introducing the relevant codes and standards, as well as the topical serviceability limit states and ultimate limit states. These design aspects cover steel pipes, including yielding, buckling, bursting and fatigue, pipeline stability on the seabed, including free spanning, expansion and buckling, as well as corrosion prevention by coating and cathodic protection.

Chapter 7 deals with the fabrication of individual pipe joints, which includes linepipe production, sacrificial anode manufacture, anti-corrosion coating and insulation, anode installation and concrete coating. Pipeline installation in shallow or deep water is treated in Chapter 8. This involves assembly of the joints into pipe strings, which are installed by pipelaying, reeling, towing or directional drilling, followed by trenching and backfilling, as required. Separate sections describe crossings, shore approaches, riser installations and tie-ins, and pre-commissioning activities. The requirements to quality control and documentation throughout the design and installation process are reviewed in Chapter 9.

Fittings and other components, such as flanges, tees, valves and other appurtenances are included in the chapters mentioned above, whereas the special conditions related to flexible pipelines are treated separately in Chapter 10. In addition to a description of the flexible pipe structure, this involves a complete overview of design, fabrication and installation, as well as testing and qualification of materials and prototypes. The final chapter deals with the operation of pipeline systems, including commissioning, inspection, maintenance and repair.

The Glossary and notation (see pages xvi–xlii) explains more than 200 special terms as they are used in the text, defines all acronyms (referring to a relevant section for further information) and identifies the frequently used symbols. Unless otherwise indicated, SI units are used throughout, and all formulae are dimensionally correct.

The volume is concluded by a Bibliography and References, divided into regulatory documents (codes, standards, recommendations, guidelines) and other references (monographs, papers, reports) cited in the text. The Index is divided into a Subject Index, listing pages giving information on the subject matter, and an Author Index, listing pages where each cited author is mentioned.

Chapter 2
Bases for design

2.1 Introduction

The objective of a marine pipeline is to transport a medium from one location to another. Many different parameters – economic, technical, environmental, etc. – determine whether or not a marine pipeline system will be installed. The justification may not rely solely on assessments of cost estimates and transportation requirements. Decisions may also be influenced by technically less tangible aspects such as societal expectations of security of supply, requiring sufficient redundancy in pipeline networks, or the political objectives of opening up new oil or gas provinces for economic or strategic reasons. In most cases the needs for transportation will be fairly well defined, although a degree of uncertainty as to quantities produced or projected hydrocarbon sales will normally be present. Even with a fair degree of certainty concerning throughput requirements, the determination of basic design parameters such as design pressure and pipeline size can imply considerable engineering. This is particularly true for large gas transmission lines.

The bases for design consist of the basic requirements to functionality, as well as a description of the environment into which the pipeline will be placed, leading to the selection of pipeline dimension and routing. A large number of requirements may be included in the bases for design. These comprise the physical pipe properties, such as diameter, steel grade options and linepipe specification details, including supplementary requirements to codes and guidelines. Central also is the definition of parameters regarding flow assurance and pressure containment, i.e. design temperature and pressure, maximum and minimum operating temperature, maximum operating pressure, and details of incidental operation. Other factors include corrosion allowance, sweet or sour service, pipeline protection principles, and possibly a number of design philosophy statements, where the use of proven technology is often specified.

A Design Basis or Pipeline Design Book would be a suitable reporting format for the bases for design; see Chapter 9. The alignment sheets (see Section 2.6.2) document not only the physical environment (features of the selected route), but also the outcome of the design activities, being the main vehicle for information from the owner's design team to the installation contractor.

2.2   Basic requirements

2.2.1   *Functional requirements*

The primary function of a pipeline is to transport the medium safely and reliably during the planned service life. The service conditions for the pipeline are related to a medium with elevated pressure, flowing at temperatures that will vary along the route from typically a high inlet temperature to temperatures that may be critically low. In gas pipelines low temperatures may cause the formation of hydrates, while in oil pipelines waxing and viscosity problems may arise.

The functional or operational requirements basically concern the operation of the pipeline. The requirements cover definitions of the system's ability to transport a specified medium quantity within a temperature envelope, as described in Section 2.3 below. The requirements also relate to the service and maintenance of the pipeline system. Other requirements may arise from safety assessment or operator practice, and imply the introduction of subsea isolation valves, monitoring systems, diverless access and so forth. Functional requirements also include the requirements to inspection access, normally pig launchers and receivers. For pipelines terminating on manned platforms or terminals the integration with fire fighting and other safety systems falls under the heading of functional requirements.

2.2.2   *Authorities' requirements*

When drafting the project parameters, including the bases for design, it is important to evaluate carefully the organisation and scope for the authority engineering. Getting consent from authorities can be a surprisingly prolonged affair, and unless thoroughly planned the overall contract schedule could have the authorities' approval on the critical path, which coupled with the sheer complexity of the approval procedures could lead to a less than cost optimised construction. The recommendation is to allow sufficient time and resources for authority engineering from the very start of the pre-engineering phase.

The authorities involved would normally include energy agencies, naval authorities, environmental agencies, natural resource agencies, health and safety bodies, work authorisation authorities, and various regional and national nature protection agencies, particularly when landfall construction is included. For transboundary projects, typically large gas transmission lines, the authorities' approval becomes increasingly complex.

Particular complications may arise when during the approval process other users of the sea claim that they will incur either temporary or in some instances permanent loss of income following the construction of a pipeline. Authorities will listen to the parties and normally secure due process. In particular, fishing organisations in many countries are very vociferous in defending their interests. There are regional differences and therefore the conditions at the pipeline location must be examined. As an example, Denmark has a well structured agency based system representing other parties, whereas some other countries do not. Even in a mature oil and gas producing area such as the UK, no such structured system exists, and there seem to be changes on a project to project basis. The establishment of agreed procedures for handling third party claims should therefore be given high priority.

A recommendation to project owners would therefore be to take very seriously the task of processing permissions and approvals for everything from survey operations to construction vessels with four-berth cabins. From an objective point of view it is indisputable that a large diameter gas pipeline, passing a sea area used by or of potential interest to others, requires careful planning. Therefore, in most countries authorities may influence the framework of the bases for design.

2.2.3 *Environmental impact*

Marine pipeline projects seem increasingly to be covered by national authority regulations requiring environmental impact assessment (EIA). Though an EIA has not traditionally been mandatory for offshore interfield pipelines, it is most often the case for pipelines approaching shores. There are profound reasons for taking this activity seriously. There are countries in Northern Europe where the time required to undertake a full environmental impact assessment (EIA) is approximately two years. For scheduling purposes this is of obvious importance.

When authorities evaluate whether an EIA is required, an often-used criterion will be whether the pipeline route is within a national economic near shore zone, the 12 nautical mile zone. Another criterion will be if the project includes landfalls, in which case an EIA will normally be required. However, no general guidelines exist, and the evaluation will therefore be country specific.

EIA obligations in general follow regional and international treaties, the most widely known being the international UNECE Convention on Environmental Impact Assessment in a Transboundary Context, signed in Espoo, Finland in 1991. The Espoo convention, which came into force in 1997, stipulates the obligations of parties to assess the environmental impact of certain activities at an early stage of planning. It also lays down the general obligation of states to notify and consult each other on all major projects under consideration that are likely to have a significant adverse environmental impact across boundaries. This convention can be flagged by relevant authorities or interested parties in cases where pipelines cross national territorial sea boundaries.

Within the European Union, EU Council Directive 97/11/EC (amending Directive 85/337/EEC) covers environmental impact assessment for certain public and private projects. The implication here is that mandatory requirements do exist for certain large diameter pipelines.

Some guidance for performing EIA can be found in *Environmental Guidelines for Selected Industrial and Power Development Projects* from the Asian Development Bank (1990) and in *Environmental Assessment Sourcebook, Volume III* from the World Bank (1991).

Typically an EIA report includes headlines such as:

- background and legal framework for EIA;
- description of the project and project elements;
- description of construction activities and schedules;
- description of operational phases;
- possible environmental impacts of construction and operations;
- mitigating factors and activities;
- monitoring requirements;
- omissions in the EIA.

Possible impacts from pipeline construction and operation in the coastal and landfall zone area are listed in Table 2.1, together with mitigating actions.

It follows from the above brief review that the EIA is an integrated part of the design premise, and should be given attention and manning in parallel with the more traditional activities such as design and procurement.

A method of evaluating the environmental impact of the project is to use the so-called BATNEEC (Best Available Technology Not Entailing Excessive Costs) method. The principle is that the best available technology should be chosen taking into account the cost involved. The BATNEEC method is in this way similar to the ALARP principle used in the evaluation of risk, see Section 5.4.

In all countries, at least where EIA legislation is in place and enforced, there are legal counsel who specialise in environmental approval advice. Although an apparently expensive exercise, it has been proven useful to engage legal counsel in the countries where the route passes their territorial waters, whether inside or outside the 12 mile exclusive economic zone (EEZ).

2.3 Flow calculations

2.3.1 *General*

Some of the main elements of pipeline system design are:

- the establishment of operational parameters;
- pipeline size determination;
- flow simulations.

Table 2.1 Environmental impacts from pipeline projects

Negative environmental impact	Mitigating action
Sediment dispersion, dispersion of drill mud or cuttings	Select or optimise construction method with regard to trenching, dredging or drilling Plan construction in period with minimum wave and current action
Suspension of toxic sediments	Change pipeline route and/or change construction method
Damage to benthic community	Minimise trenching and burial
Interference with fishery	Speed up construction or change construction method Optimise time schedule and keep close contact with local fishery organisations
Dispersion of pressure testing water	Minimise the use of chemical treatment, e.g. through residence time optimisation Change procedure, release water offshore
Degradation of landscape and scenic value	Optimise construction method. Make careful restoration
Degradation of recreational values	Minimise construction schedule
Increased traffic	Optimise construction schedule and activities
Increased noise and atmospheric emission	Improve equipment
Degradation of corrosion protection	Select non-toxic protection system
Accidental loss of product or chemicals	Optimise operation and maintenance

Flow simulations are an integrated part of the system design, and are required for the optimisation of the pipeline size. The simulations provide pressure and temperature profiles along the pipeline. The profiles subsequently constitute input information to the operational parameters, and are of course dependent on the pipeline dimension.

As part of the pipe size optimisation, important decisions have to be made regarding the pipeline system:

- single or dual pipeline;
- selection of pipe material (carbon steel, corrosion resistant alloy, etc.);
- pipeline coating and insulation.

2.3.2 *Operational parameters*

As a basis for the system design it is necessary to know the operational parameters for the pipeline system. Such parameters are: the amount of medium to be transported; the composition of the transported medium; the temperature of the medium, etc. The operational parameters will normally be selected as the design basis for a given product.

In the design situation it is necessary to take possible changes of the basic parameters into account. This can be relevant for the composition of the transported medium, as well as the temperature, pressure, etc.

The composition of the transported medium will determine the selection of the pipe material. Hydrocarbons containing high quantities of CO_2 or H_2S may require the use of high alloy steels (e.g. stainless steel) or clad pipe, particularly in the presence of water or elevated temperatures.

2.3.3 Pipeline size determination

The determination of the pipeline diameter is performed on the basis of the main operational parameters for the pipeline system, such as:

- annual flow;
- expected system availability (load factor);
- requirements for delivery pressure;
- properties of the transported medium.

The optimum pipeline size is based on a 'lifetime' evaluation of the system, taking into account the capital cost (CAPEX) for the establishment of compressor/ pumps, the pipeline itself, receiving facilities, as well the operational cost (OPEX) for the system.

An economic model for the pipeline system is often used to calculate different economic key parameters such as: net present value (NPV), unit transportation cost (UTC), etc., as illustrated in Figure 2.1.

An important part of the optimisation process is the flow calculations, including the requirements for compression or pumping. In the initial phase the flow calculations may be performed on an overall level without detailed modelling of the thermodynamic conditions along the pipeline. However, such modelling may eventually be required because parameters other than the pressure drop may be important factors for the pipeline dimension.

2.3.4 Flow simulations

For gas pipelines as well as for oil pipelines thermodynamic simulations have to be performed to evaluate the conditions of the flowing medium. For gas pipelines it is necessary to analyse the possibility of liquid drop-out, of both water and hydrocarbons, as well as hydrate formation. For oil pipelines it is wax deposition or two-phase (gas–liquid) flow that are of concern. The likelihood and degree of internal corrosion and erosion are also closely related to the flow conditions. The

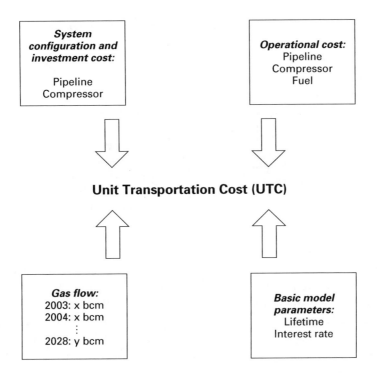

Figure 2.1   Illustration of cost optimisation model for determination of pipeline size

principal issues are briefly identified below. For a more comprehensive discussion of flow assurance reference is made to Section 11.2.

Pressure and temperature profiles

To determine the temperature and pressure profiles, the thermal interaction with the surrounding soil and water should be established. The heat transfer will depend on the pipeline material, wall thickness, insulation medium, fluid temperature and properties, the burial depth, type of soil, etc. In most cases computer programs are used to simulate the flow conditions in the pipeline. The required detail and specific problems in relation to the flow calculations are highly dependent on the transported medium.

Note that for flowlines carrying hot well fluids cooling may be required to obtain acceptable outlet temperatures. In such cases burial of the pipeline may have to be avoided.

For natural gas it is necessary to take into account the Joule–Thomson effect, whereby the gas will cool due to its expansion as the pressure decreases along the pipeline. The temperature drop will be related to the pressure drop in the pipeline, and minimum allowable temperatures may need to be calculated to determine the capacity of the pipeline. Joule–Thomson cooling is proportional to the slope of the pressure profile curve, which in most cases will increase in the downstream end of the pipeline. The gas temperature during normal operation may drop below zero, resulting in permanent freezing and ice formation around the pipeline. Repeated freezing and thawing cycles imply a risk for uplift of the pipeline.

In addition to the evaluation of the flow conditions in the pipeline system, the pressure and temperature profiles will provide input to a number of other engineering studies in relation to the pipeline design, such as:

- sacrificial anode design;
- pipeline expansion;
- free span evaluation;
- trawl impact resistance.

Liquid dropout

In the case of two-phase (oil and gas) flow, slugs may form, particularly at low points in areas with uneven seabed. Drop-out of water and natural gas liquids (NGL or condensate) can occur in gas pipelines at certain pressure and temperature conditions. If slug formation is likely, the pipeline system should be designed for such slugs, typically a slug catcher is sized and located at the receiver terminal. In most cases, evaluation of slug sizes requires dynamic flow simulations for the pipeline system. Alternatively, the design process can consider operating outside the regions where slug formation is problematic.

Wax deposits

Wax deposits may occur in pipelines transporting oil. The wax will reduce the free bore of the pipeline and thus the capacity of the pipeline system. The wax deposits will begin to form when the medium reaches a certain low temperature, and it may be necessary to insulate the pipeline to avoid such wax formation. The effects of wax deposits may be reduced by regular pigging of the pipeline.

Gas hydrates

The presence of free water in a gas pipeline may result in the formation of gas hydrates, which are ice-like deposits that are formed under certain temperature and pressure conditions, depending on the composition of the gas. The hydrates have a strong tendency to agglomerate and stick to the pipe wall, thereby plugging up the pipeline. Gas hydrate deposits may thus reduce the possible throughput, or even result in total blockage of the pipeline.

Hydrates occur when a gas stream is cooled below its hydrate formation temperature, i.e. when the gas temperature is below its water dew point temperature, leading to water drop-out. Hydrate formation should be avoided because the hydrates do not dissociate under the same conditions as those at which they were created. Significantly higher temperatures and/or lower pressures are required and, even at these conditions, hydrate dissociation is a slow process.

The hydrate prevention philosophy is based on three levels of security, listed in order of priority.

(1) Avoid operational conditions that might cause formation of hydrates.
(2) Prevent the formation of hydrates by adding chemicals that lower the threshold of the hydrate formation (inhibitors).
(3) Temporarily change operating conditions in order to avoid hydrate formation.

Corrosion and erosion

The risk of corrosion and erosion of the pipeline is closely related to the flow conditions. The corrosion risk is related to the presence of water and corrosive components such as CO_2 (sweet corrosion) and H_2S (sour corrosion). Erosion in the pipeline is due to sand and high fluid velocities, and the risk of erosion damage is most acute for bends in the pipeline and control components such as chokes and valves.

2.4   Site investigations

2.4.1   *General*

Geotechnical data and design information are established through site investigations. The objective is to obtain reliable data for the design, construction and subsequent operation and maintenance of the pipeline and associated structures. All geotechnical design parameters along the entire pipe route should be determined to a depth where the soil strength and flexibility have no influence on the pipeline.

Site investigations for pipeline projects comprise geophysical as well as geotechnical surveys. The survey activities are normally performed in phases, each phase adding information to the previously established data, or improving the reliability of existing data. In this way the investigations can be integrated with the engineering and construction activities, and the extent and level of detail of the site investigations can be optimised according to the project requirements. An overview of the site investigations is presented in Figure 2.2.

It may be useful to outline the main difference between geophysical and geotechnical investigations, both of which have the objective of describing the properties of the earth. Geophysics is the non-invasive investigation of subsurface conditions in the earth through measuring, analysing and interpreting physical fields at the surface, whereas geotechnical investigations are based on indirect measurements of soil properties, either in situ or on soil samples in the laboratory.

Geophysical investigations typically apply seismic reflection and refraction, gravity, magnetic, electrical, electromagnetic and radioactivity methods. Some

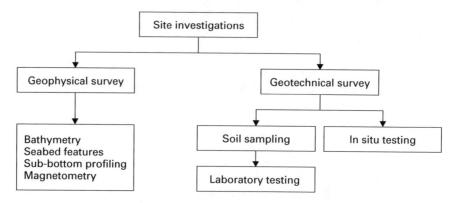

Figure 2.2   Schematic illustration of site investigations

studies are used to determine what is directly below the surface (the first metre or so), other investigations extend much deeper.

2.4.2   *Geophysical survey*

A geophysical survey is performed from a survey vessel equipped with a number of sensors and data gathering systems that are tailored to suit the specific soil conditions and the requirements of the particular pipeline project. In general, the equipment includes as a minimum:

- a precision surface position system;
- an echo sounder, and/or multi-beam echo sounding system;
- a side scan sonar;
- a seismic profiler;
- a computer based integrated on-line analysis and presentation system.

In shallow water the subsea sensors may be hull-mounted, whereas in deeper waters the sensor carrier may be a towed fish or a remotely operated vehicle (ROV).

Echo sounder

The echo sounder provides seabed profile and bathymetric mapping along the pipeline route. The width of the survey corridor is typically between 500 m and 1000 m, centred on the proposed pipeline route. The width of the pipeline corridor depends on the topical water depth. In shallow areas and at shore approaches 500 m may be sufficient, whereas 1000 m is more appropriate in deeper waters, even with enhanced precision requirements to the acoustic system. The increase in corridor width is required partly because the accuracy of the acoustic measuring system decreases with increasing water depth, and partly because pipeline installation in deeper waters is associated with a larger uncertainty.

Side scan sonar

A plan view of the seabed is provided by a side scan sonar, which can produce slant-corrected mosaics of consistent scale. It gives an almost photographic view, and seabed features, such as rock outcrops, debris or wrecks can be identified very clearly.

Seismic profiler

Seismic profiling produces vertical profiles of the seafloor and the strata below from the reflections generated from changes in acoustic velocity and impedance. They give a clear picture of geological features, such as filled in depressions or valleys, rock ridges, soil containing boulders or gas. Interpretation together with an analysis of in situ testing can give a continuous description of the top soil strata along the pipeline route. The information gained is qualitative rather than quantitative. The required vertical resolution should be specified, because the depth of interest is limited, and far less than that required for platform foundations or the seismic investigations for oil exploration.

Magnetometer

The total intensity of the Earth's magnetic field is detected and recorded by a magnetometer, which is designed to be sensitive to local variations in the magnetic field. This makes it useful for locating pipelines, armoured cables, wellheads, wrecks and other ferrous-metal objects on or just below the seabed.

2.4.3   Geotechnical survey

A geotechnical survey is required for soil sampling and in situ testing of soil properties. The soil properties are required for classification and design parameters, which can be used for interpretation, correlation and calibration of the geophysical survey. The geotechnical survey can be performed using the same type of vessel that was mobilised for the geophysical survey as long as samples and testing of the seabed surface soils are required (0–5 m below seabed surface). Soil samples from a greater depth (dependent on soil strength) require the application of a vessel capable of geotechnical coring.

Geotechnical testing and sampling should be performed at regular intervals along the pipeline route. The spacing has to be selected based on the project needs and the expected variability along the route. Typical distances between testing or sampling is in the range 500 to 5000 m.

The spacing of the geotechnical sampling is in most cases larger than accepted for traditional onshore projects, thus in the interpretation of the soil conditions from geotechnical samples there can be a large degree of uncertainty. It is therefore important to link the geophysical and the geotechnical survey information, so that the geophysical data can be calibrated at the available sampling locations, and can support the interpretation of soil conditions between the samples.

The *Guidance notes on geotechnical investigations for marine pipelines* prepared by the Pipeline Working Group of the Offshore Soil Investigation Forum provides guidance on good practice in the collection of geotechnical data, and aid in planning and specification of marine pipeline geotechnical surveys. A brief review of geotechnical soil sampling, in situ testing and laboratory testing is given in the following sections.

2.4.4 *Soil sampling and in situ testing*

Site specific soils information is obtained through soil sampling and in situ testing. This may be performed as part of the geophysical survey or may be carried out through a dedicated geotechnical survey. The soil investigations are performed using the following equipment:

- grab sampler;
- vibrocorer;
- gravity/drop corer;
- cone penetrometer;
- vane shear tester.

Grab sampling
A grab sampler is capable of retrieving a superficial sample of disturbed seabed material. In view of the limited depth and the disturbed nature of the sample this type of sampling has limited application.

Vibrocoring
A vibrocorer is self-powered and capable of obtaining soil samples several metres long. The driving force in the tool comes from an electrical vibrator motor, driven through electrical cables supplied from the surface vessel. Vibrocoring is used in hard or dense material where gravity or drop corers may have very little penetration.

Gravity coring
The gravity corer or drop corer consists of a weighted coring tube, which relies on gravity for seabed penetration. The tube is 3 to 6 m long, and shaped in the form of a dart. The penetration into the seabed depends on the strength of the soil and the weight of the corer. The penetration is in general good in low to medium strength clays and low in hard clay or sand.

Soil testing
The cone penetrometer and the shear vane are used to determine the in situ soil properties. The cone penetrometer records the soil resistance when an instrumented cone is forced into the seabed. In order to improve the reliability of the prediction

Table 2.2 Standard soil test for pipeline projects

Test type	Application
Sieve test, particle size distribution	Classification of sand and silt Sediment transport/scour considerations
Atterberg limits	Classification of clay
Water content, void ratio, bulk and dry density, specific gravity	Bearing capacity, liquefaction, on-bottom stability
Triaxial compression test	Bearing capacity, on-bottom stability, trenching
Shear box test	On-bottom stability, trench slope angles
Consolidation test	Settlement
Specific resistivity test	Cathodic protection design

of soil properties based on the cone penetration, core samples should be collected at a number of locations for cone penetrometer testing (CPT). In clays the vane shear tester gives very good measurement of the undrained shear strength.

2.4.5 *Laboratory testing*

A full laboratory testing programme should be performed on extracted soil samples. Such a programme includes standard classification and strength and deformation tests for determination of engineering parameters. Standard tests and their application in pipeline projects are summarised in Table 2.2.

The reporting of the testing should include descriptions of methods and equipment, evaluation of results, and presentation of the geological model along the alignment.

2.5 Meteo-marine data

2.5.1 *General*

The acquisition of meteorological and hydrographic data of relevance for the pipeline project is traditionally referred to as environmental investigations. However, so as not to confuse the issue with the investigations of environmental impact (see Section 2.2.3 above) the term meteo-marine or metocean is increasingly used. The objective is to establish hydrographic data and parameters to be used in the design, installation and operation of the marine pipeline. The most important data describe the wave, current and water level conditions along the pipeline route. Other data that may have significant impact on the pipeline project are water temperature and salinity. Wind data are often considered useful in pipeline projects because of the close correlation between wind and waves. At many locations wind data are more readily available than wave data, and can thus increase the reliability of the wave information.

The water level is normally defined relative to some fixed reference level, e.g. Lowest Astronomical Tide (LAT). The actual water level varies due to tides, storm surges and other natural phenomena. The water level, together with the bathymetric information, defines the water depth, which is required for a number of design activities, e.g. calculation of wave and current flow at the pipeline level; see Chapter 6.

2.5.2 *Wind, waves and current*

The description of wind, waves and steady current includes the following:

for wind:

- wind generation processes;
- weather patterns generating storms;
- determination of storms;

for waves:

- wind waves and wave theories;
- statistical parameters describing waves and wave spectra;
- directionality of waves;
- effect of water depth.

for currents:

- different types of current;
- current profiles.

Success in reconciling economic design on the one hand with demonstrable safety of the pipeline on the other is strongly dependent on the availability of accurate meteo-marine data. This consideration should be made fundamental to the planning and execution of all environmental data gathering and analysis.

Based on the origin of data, namely the atmosphere and the sea, it is convenient to group the data into meteorological data and hydrographic data. Each group can be subdivided into:

- observed data;
- measured data;
- model data.

The environmental data in mature offshore areas, for example the North Sea or the Gulf of Mexico, are potentially vast and of high quality. Because the data are established by operators for specific projects the availability of the data is in most cases restricted to these operators. The major problem in these cases is to get operators to pool and share the information.

Projects located outside well explored areas are often faced with the problem of environmental data that are insufficient in quantity or of relatively poor quality. In this case two main options exist for establishing additional data:

- initiation of a field measuring programme;
- application of mathematical models.

Which option to choose will depend on several factors, such as the status of existing data, the location of the project area, the level of analysis to be applied, and the available budget and time. Often a combination of a measuring programme and mathematical models will establish the best basis.

2.5.3   *Collection of wave data*

The site specific wave conditions may have a significant impact on pipeline projects, technically as well as cost-wise. The wave conditions influence the pipeline installation method, the choice of pipelay vessel, and the feasible installation period. On water depths less than 100–200 m, waves are furthermore paramount for the selection of the minimum submerged weight of the pipeline. Wave data are therefore often measured at the initiation of pipeline projects in new areas in order to improve the reliability of the knowledge of wave conditions, and thereby reduce project uncertainties. A brief review of methods for wave measurements is therefore presented below.

Waves can be measured using a variety of instruments: surface buoys, acoustic current doppler current meters, platform mounted instruments such as laser and radar altimeters, or satellite borne instruments.

Wave rider buoys
Wave rider buoys are reliable and accurate for the determination of the significant wave height and mean or the peak period of the spectrum. The inherent properties of the buoys imply that the wave crests of maximum waves and the steepness of the wave front is determined less accurately.

Doppler sonar current meters
Wave measurements can be collected by doppler sonar current meters, using particular software for analysis of the recordings. Conventional acoustic doppler current profilers (ADCPs) typically use four acoustic beams paired in orthogonal planes; each beam is inclined at a fixed angle to the vertical. The sonar measures

the component of the velocity projected along the beam axis, in cells approximately half the length of the acoustic pulse in the direction of the beam. Determining the mean horizontal current from these data is straightforward. The situation regarding waves is more complicated because the wave velocity varies across the array at any instant of time. As a result it is not possible to decompose the along-beam velocity into horizontal and vertical components. However, the wave field is statistically steady in time and homogeneous in space so that cross spectra of velocities measured at various range cells depend on the wave direction. This makes it possible to apply array-processing techniques to estimate the frequency–direction spectrum of the waves.

Radar and laser altimeters

Mounted on fixed platforms, radar and laser altimeters give very reliable data in most cases. One drawback is that under severe storm conditions the meter readings may be hampered by sea spray.

Satellite borne sensors

Satellite borne sensors for wave measurements exist in the form of radar altimeters and Synthetic Aperture Radar (SAR). The radar altimeter can in principle only give the significant wave height, but the period can be estimated on detection of the wave slope. SAR provides a full directional spectrum derived from an image of the sea surface. SAR wave data are particularly valuable for the assessment of wave conditions in areas where no other measurements exist. The data can provide operational information or can in combination with numerical hindcast models provide design data. The main applications of the satellite data are calibration and validation of the numerical wave models.

2.5.4 *Design parameters*

Various general and national design regulations specify that offshore structures should be designed to withstand the statistical T-year maximum environmental design load, where the return period, T, is normally 100 years. However, the same regulations rarely specify the methods to be applied for the calculation of the T-year, or the degree of confidence that it should be possible to attribute to the resulting design parameters.

Basically the same methods can be applied for calculating extreme values of any type of environmental parameter, such as wave height, current velocity, wind speed or water level. The main tasks of such statistical analyses are as follows.

- Check that the data are independent and homogeneous. Possible seasonal variations have to be considered, for example, the analysis of typhoon data has to be performed independently of monsoon data.
- Select the analysis approach: either annual maximum method or peak over threshold method.

- Select candidate distribution or distributions.
- Fit the data to the distributions, and check the quality of the fit.
- Extrapolate to the T-year event.
- Estimate uncertainty with the T-year event.
- Consider joint probability of different parameters.

Joint probability is an important aspect, which in most cases is difficult to incorporate accurately. It is rare that the data allow determination of unambiguous joint probability, so extrapolations are made independently for each parameter. The combination of the different extreme parameters is thereafter performed by simple, but conservative methods, see Chapter 6.

Design data describe global parameters, whereas the design activities in most cases require information on the data in the immediate vicinity of the pipeline. Global data must therefore be transferred to pipeline level using different theories or statistical methods. Global wave data are transferred to the pipe level using wave theories, which may be more or less complex. First-order wave theory is in many cases adequate, but higher-order theories or wave directionality may be required on shallow to medium water depths.

2.6   Route selection

2.6.1   *General*

One of the most important engineering activities, which can have a significant influence on the overall cost for the installation and operation of a pipeline system, is the definition of the pipeline route.

The pipeline route selection should be performed to give the economically optimal solution for the pipeline owner. This comprises the costs of fabrication, installation, operation and decommissioning. Normally the most cost effective solution will be the shortest possible route. However, different features along the pipeline route, such as severe seabed conditions, environmentally sensitive areas, historical or archaeological sites, existing facilities for oil/gas production, wind farms, heavy traffic shipping lanes, etc. may force the pipeline away from the most direct route.

One example is where authorities may not permit the pipeline to pass habitat areas for certain species. Other types of authorities' requirements can be that the pipeline is not allowed to be installed during certain times of the year. Such requirements should be evaluated in the pipeline route selection and the determination of the overall pipeline installation logistics.

Detailed route selection should be performed to reduce the number of free spans, and consequently the number of pipeline supports, in particular in areas with extremely uneven seabed. In such areas pipeline optimisation can be performed during pre-survey, using 3D-pipeline route optimisation software.

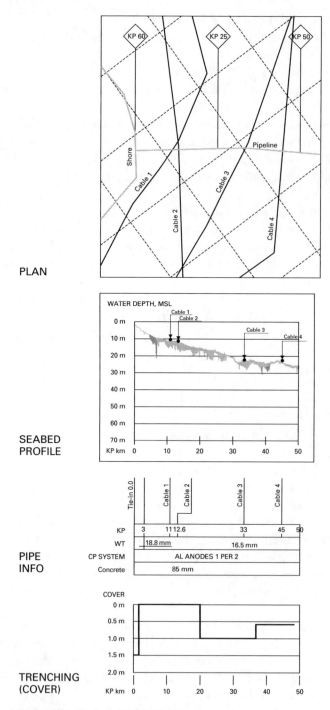

PLAN

SEABED
PROFILE

PIPE
INFO

TRENCHING
(COVER)

Figure 2.3   Typical alignment sheet

If the pipeline area is not well known to the design team it may be necessary to carry out a feasibility survey, to obtain the information required for the route selection. Normally a preliminary route selection will be performed as a starting point for the conceptual/detailed design phase. This route will be based on existing information from the pipeline area, and as part of the preliminary route selection the requirements for addition information are identified.

2.6.2 *Alignment sheets*

Normally the selected pipeline route will be documented in overall route layout drawings and so-called alignment sheets, including description of seabed soil conditions and other types of useful information about pipeline features.

The route layout drawings would include a bathymetric seabed profile and a plan of the pipeline, the length being specified in a kilometre post (KP) system. The plan should also identify straight and curved (circular) sections, separated by tangent points (TP). A TP list defines the corresponding KP and geographical coordinates, as well as the coordinates of the tangent intersection point (IP) and the circle centre point (CP) for each curved section.

In the alignment sheets additional information such as cable or pipeline crossings, required soil cover, sacrificial anodes, etc. is given. Figure 2.3 shows a typical alignment sheet, covering a substantial pipeline length. Detailed alignment sheets, issued to the installation contractor, would typically cover a length of 2.5 km, and contain information on:

- linepipe steel grade, ID and WT;
- anti-corrosion coating thickness and density;
- concrete coating thickness and density;
- field joint coating type and infill density;
- sacrificial anode type and spacing;
- pipe weight, in air and submerged;
- buckle arrestor type and spacing;
- seabed intervention works;
- maximum pipelay tension;
- pipelay tolerances.

The plan drawing may incorporate contour maps from the survey, as well as identification of kilometre posts, tangent points, curvatures, crossings, pipeline fittings (tees, wyes) and seabed features (boulders, wrecks).

The alignment sheet will be the basis for determining the pipeline route length, and as such the basis for bill of quantities, or material take-off (MTO) for the tendering of linepipe and anodes.

Chapter 3
Materials

3.1 Introduction

The material parameters that are relevant for design and installation of marine pipelines are on the one hand those of the seabed on which the pipeline is placed, and on the other hand those of the constituents making up the pipeline itself. The seabed – whether natural strata or engineered backfill – is part of the environment, and is described in Chapter 2, while the properties are treated in this chapter, along with the physical properties of the pipeline. Only the material characteristics of the latter are discussed here, the manufacturing processes being described in Chapter 7.

The seabed materials are described by a number of geotechnical parameters, and their properties are described by strength and deformation parameters. The primary constituent of the pipeline is the linepipe material, which is principally characterised by the mechanical properties, and by the resistance to deterioration mechanisms caused by the external environment or by the transported product.

The primary measure that can be taken against external corrosion is cathodic protection by sacrificial anodes, and its effectiveness depends upon the electrochemical properties of the anode materials. Cathodic protection is usually used in combination with some form of external coating. The external coating may also be used for thermal insulation, as well as mechanical protection.

Internal corrosion is mitigated either by controlling the internal environment (e.g. by removal of water by drying, use of specific corrosion inhibitors, frequent pigging, etc.), by use of corrosion allowances in the pipewall thickness or by use of corrosion resistant alloys. Corrosion resistant alloys may be used as solid pipes or as a relatively thin (usually 2–3 mm) cladding, lining or weld-overlay at the internal pipe surface.

Coatings may also find use as internal corrosion protection, but they are normally only intended to reduce friction and to prevent internal rusting prior to the construction phase and during commissioning. Special internal coating systems have been developed to protect the pipe joint zones (girth welds) internally, but they have not gained general recognition as reliable solutions.

3.2   Soil materials

3.2.1   *General*

A classification of the soil is required to determine the engineering properties that are applied in design. It is primarily strength and friction properties that are of concern in pipe–soil interaction. Typical strength parameters for cohesive and non-cohesive soils are given in Tables 3.1 and 3.2. The engineering activities dealing with pipe–soil interaction can be grouped under the following main headings:

- vertical stability;
- settlement;
- flotation (liquefaction);
- lateral resistance to pipeline movements;
- axial resistance to pipeline movements;
- scour and/or self-lowering;
- free span evaluation.

Recommended specific soil parameters to be used in the engineering activities related to the above headings are given in Section 6.3.6.

3.2.2   *Seabed soil classification*

The classification of the soil is based on visual inspection and laboratory testing. The aim of the investigation is to classify the soil so that the strength parameters can be determined. The visual description should include:

Table 3.1   Cohesive soil consistency classification

Consistency	Shear strength (kN/m^2)
Very soft	0–12
Soft	12–25
Firm	25–50
Stiff	50–100
Very stiff	>100

Table 3.2   Non-cohesive soil characterisation

Compactness	Very loose	Loose	Medium dense	Dense	Very dense
Relative density (%)	0–15	15–35	35–65	65–85	85–100
Unit weight, moist (kN/m^3)	<16.0	15.0–20.0	17.5–20.5	17.5–22.5	>20.5
Unit weight, submerged (kN/m^3)	<10	8.8–10.0	9.4–10.5	10.5–12.5	>10.5
Friction angle (°)	<28	28–30	30–36	36–41	>41

- main soil component;
- consistency, colour, content of organic materials;
- texture referring to grain size.

Laboratory classification tests include:

- grain size and distribution;
- density and water content;
- strength and plasticity.

Grain size and distribution is found through a sieve analysis. Distribution of fine grained soil components (silt, clay) is found by hydrometer analysis, which is applied to soil materials passing the No. 200 (0.074 mm) sieve. The analysis measures the fall velocity of the fine grains in aquatic suspension, which depends upon the grain size, and letting the suspension settle for a given time establishes the grain size distribution. The hydrometer is scaled to give the specific gravity of the suspension or the grams per litre of suspension, see for example ASTM D422-63 *Standard test method for particle-size analysis of soils*.

In cohesionless soils the relative density is useful for characterising the soil. The relative density ρ_r is defined as:

$$\rho_r = (e_{max} - e)/(e_{max} - e_{min})$$

where

e_{max}	void ratio of the soil in its loosest state
e	in situ void ratio
e_{min}	void ratio of the soil in its densest state.

Non-cohesive soils can be characterised according to Table 3.2.

The plasticity characterises the states of the soil in four categories depending on the water content:

- brittle solid;
- semisolid;
- plastic solid;
- liquid.

The water content at the boundary between semisolid and plastic solid is defined as the plastic limit (*PL*), and the water content at the boundary between plastic solid and liquid is defined as the liquid limit (*LL*). Tests for determining these limits were originally proposed in 1911, by Mr Albert Atterberg. The tests (Atterberg Limits) have been modified and combined into a standard test procedure as, for example, outlined by ASTM D4318.

The plasticity index (PI) is defined as:

$$PI = LL - PL$$

PI is particularly useful for fine-grained soils, for which it has been found that a number of engineering properties correlate well with PI. For this reason a liquidity index has been defined as:

$$LI = (WC - PL)/PI$$

where

WC in situ water content of the soil.

The water content is found by dividing the weight of the water by the weight of the solid, and is expressed in percent. If $LI < 1.0$ the soil behaves as a plastic solid. If $LI > 1.0$ the soil behaves as a viscous liquid when sheared.

3.2.3 *Backfill materials*

Engineered soil materials may be used for modification of the seabed, see Section 8.2, and in the case of pre-lay intervention these materials replace the natural seabed as the foundation for the pipeline. Soil replacement to improve the foundation properties would typically employ dredged sand, whereas crushed rock is commonly used for purposes of support or protection.

Rock materials are characterised by the median grain size D_{50} and the maximum grain size D_{max}. Grains of less than 60 mm are termed gravel, whereas those with larger dimensions are referred to as cobbles. The size required for stability on the seabed depends upon the topical wave and current action, but for deep water application with moderate currents a typical specification would be $D_{50} = 50$ mm and $D_{max} = 125$ mm. The important engineering properties are the bulk dry density and the submerged unit weight, which depend upon the grain density and the void ratio. For the typical rock materials referred to above the dry density is in the range 1.4–1.7 t/m^3 and the submerged unit weight is in the order of 10 kN/m^3.

3.3 Linepipe materials

3.3.1 *General*

The pipelines of the early 20th century were typically small diameter, seamless steel pipes operating at low pressures. As welding technology evolved, pipes were also made as longitudinally or spirally seam welded cylinders produced by forming of plates or strips. The steadily growing demand on the transport of primary energy sources such as oil and gas necessitated larger diameters and higher operating pressures. To reduce the amount of steel in the pipes there was a search for higher strength materials that could be used to reduce the wall thickness, as permitted by advances in manufacturing processes. This again enforced more strict requirements on the mechanical properties, especially in terms of preventing crack initiation and unstable crack growth, in order to reduce the associated risk of pipeline failures. Pipeline projects in critical areas such as permafrost zones and harsh offshore regions also added to the increasing demand on steel quality, as did the special hydrogen embrittlement related problems associated with the transport of sour oil and gas media. The response of the linepipe industry has been a tremendous development in steel manu-facturing, steel processing and pipe production methods to meet project-specific demands on steel chemistry, mechanical properties, weldability, non-destructive testing, geometrical tolerances, etc. The result of this refinement is generally referred to as Thermo-Mechanical Controlled Processing (TMCP), which quite often is just (imprecisely) referred to as controlled rolling. Basically, it consists of the micro-alloying of the steel to prevent unwanted grain growth during hot forming (rolling), and continuing the rolling process to lower temperatures to gain benefit from both deformation hardening and precipitation hardening.

The development from the early plain carbon steel pipes to the high-end TMCP steel pipes has provided:

- improved strength, both yield and ultimate tensile strength;
- improved toughness properties, i.e. the lowering of transition temperature from brittle to ductile fracture and an increase of the impact toughness;
- improved weldability;
- improved resistance towards hydrogen related disintegration in sour service, i.e. due to exposure to wet H_2S containing environment.

The steel may have as good mechanical properties as possible, but one thing is always inherently lacking in carbon steels: corrosion resistance. The outside of a pipeline may be protected by coating and cathodic protection as discussed in Chapter 6, but in the case of transport of corrosive commodities there is a need for corrosion resistant materials. Such a requirement is being met by the development of linepipe materials such as weldable martensitic stainless steel (chromium steels), duplex stainless steels, austenitic stainless steels, and super austenitic stainless steels and nickel alloys. The materials are generally referred to as corrosion resistant alloys (CRA), and are available either in solid form or in the form of internal lining, cladding or weld-overlay in carbon steel pipes.

Table 3.3 Increase of yield strength and ultimate tensile strength by alloying (Heller 1966)

Alloying element	Unit element (weight %)	Increase per unit element	
		Yield strength (MPa)	Tensile strength (MPa)
C	0.1	27.5	68.7
P	0.1	53.9	45.1
Si	1.0	54.9	90.2
Mn	1.0	82.4	78.5
Cr	1.0	53.9	72.6
Cu	1.0	78.5	55.9
Ni	1.0	44.1	33.4

3.3.2 *Strength, toughness and weldability*

Traditionally, alloying and heat treatment provided strength improvements. The primarily used elements were carbon (C) and manganese (Mn) but, as shown in Table 3.3, phosphorus (P), silicon (Si), chromium (Cr), copper (Cu) and nickel (Ni) influence also the strength considerably. However, there is a significant penalty in decreased weldability and loss of toughness with most of these elements, especially the C, P and Si, so the strength improvement route of alloying these elements is limited. A typical, modern linepipe steel will have a carbon content of between 0.10 and 0.15%, a manganese content of from 0.80 to 1.60%, less than 0.40% silicon, phosphorus and sulphur below 0.020% and 0.010%, respectively, and less than 0.5% copper, nickel and chromium.

The introduction of small amounts of strong carbo-nitride formers such as niobium (Nb) and vanadium (V), in addition to aluminium (Al) and titanium (Ti), facilitates grain refinement and precipitation hardening. They are all very effective even in small amounts, less than 0.01%, and steels with these elements are referred to as micro-alloyed materials.

The micro-alloying constituents are not only effective through heat treatments such as quench and tempering, but also in the thermo-mechanically processing of steel. Correctly applied, the carbo-nitride formers retard crystal growth during rolling to allow the accumulation of deformation in the austenite grains prior to transformation, and they act as seeds for ferrite formation. One of the unfortunate properties of steel is its tendency to become brittle at low temperatures. Before the introduction of microalloying, carbon steels exhibited a significant shift from normal ductile behaviour to extreme brittleness within a very narrow temperature range (usually subzero). The ductile or brittle behaviour can be measured by impact toughness testing. A pre-notched sample is exposed to impact loading (a swinging hammer), and the energy absorbed by the crack initiation and propagation is measured (by the length of travel of the hammer after impact). A brittle fracture absorbs very little energy as the cracks propagate faster than the impact loading. In a ductile material, on the other hand, the crack progress is by ductile

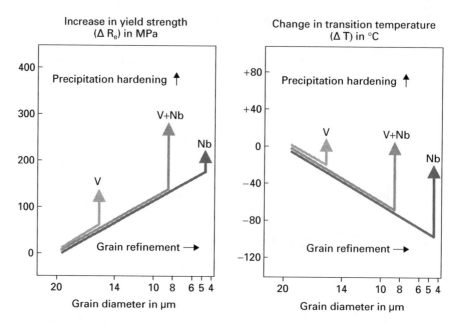

Figure 3.1   Effect of vanadium and niobium on yield strength and transition temperature of low carbon steel (Krass *et al.* 1979)

tear, which slows down the speed of the swinging hammer. The temperature where a transition from ductile to brittle behaviour takes place (defined as 50% brittle fracture appearance) is referred to as the transition temperature, and the lower this temperature is, the better suited is the material for use at subzero temperatures. Determining factors in this transition are grain size and micro-structure (in terms of strength and hardness). Large grains and strengthening by traditional alloying techniques (i.e. carbon and manganese) lead to high trans-ition temperatures. If correctly applied, the micro-alloying elements are precipit-ated as sufficiently small particles that are evenly dispersed in the matrix, and the net result is an extremely fine-grained material with increased strength and lowered transition temperatures as indicated in Figure 3.1. The micro-alloying strengthening route replaces the classical strengthening measures, and steels with less than 0.10% C and virtually free of hard microstructures can be produced with excellent mechanical properties.

As shown in Figures 3.2 and 3.3 lowering the contents of carbon and sulphur (S) can increase the impact toughness values significantly. Changing the inclu-sion form and morphology by use of rare earth materials may also increase the toughness by reducing the tendency to elongation of the manganese sulphides in the rolling plane, but they have to be used with caution if the overall cleanliness of the steel is not to be decreased.

The flip side of the 'micro-alloyed TMCP' coin is an increase in yield to tensile strength ratio, and a marked anisotropy in yield strength parallel and

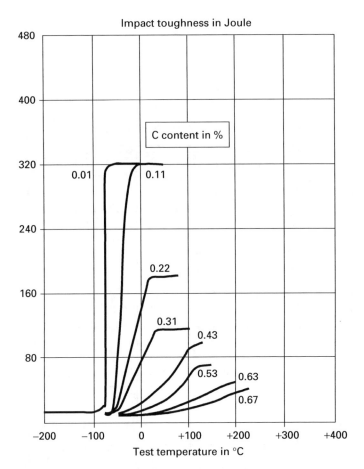

Figure 3.2   Effect of carbon on impact toughness versus test temperature (Rinebolt and Harris 1951)

transverse to the (main) rolling direction, especially in the high-end of the mechanical properties.

The primary objective in the ongoing development of steel is to prevent a fall in toughness in the weld heat-affected zone, and to control hardnesses of the weld zones in order to prevent hydrogen cold cracking (not only for the seam welds but also for subsequent field welding). The weld zone is subject to abrupt thermal cycling, and the temperatures reached in the adjacent base metal (the heat-affected zone) and the subsequent cooling rate may change the microstructure locally. This interferes with the base metal properties obtained by a delicate balance of precipitation hardening and deformation hardening by controlled rolling at lower temperatures. Small amounts of titanium (Ti) and somewhat higher amounts of molybdenum (Mo) have been effectively used to maintain the toughness in the heat-affected zones. In general, the overall improved cleanliness, lowering of the carbon content, and a balanced microchemistry in connection

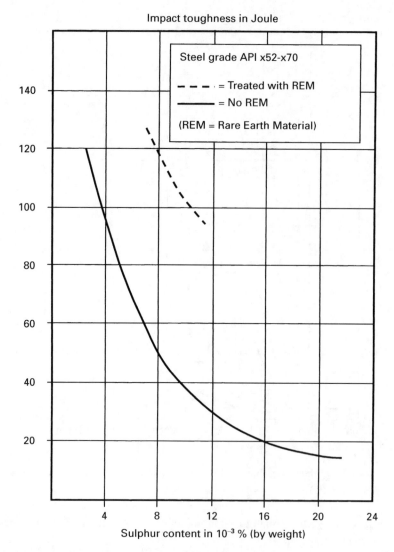

Figure 3.3 **Effect of sulphur on impact toughness of hot-rolled steel (Dahl 1977)**

with TMCP have produced high strength steel with excellent weldability. In fact the weldability problems of today are more concerned with the weld metal itself, in terms of maintaining toughness and securing overmatching mechanical properties relative to the base material.

3.3.3 *Sour service resistance*

Steel exposed to wet H_2S environments may suffer hydrogen-related damage. The various forms of hydrogen damage and their link to corrosion processes are

Simplified electrochemical reactions:

$Fe \rightarrow Fe^{++} + 2e^-$ and $H^+ + e^- \rightarrow H_{ads}$

The H_{ads} either combines to form H_2 (gas) or becomes absorbed into the steel substrate.

Presence of H_2S increases the absorption of H_{ads}, which may lead to cracking and blistering

Types of corrosion
related hydrogen
damage

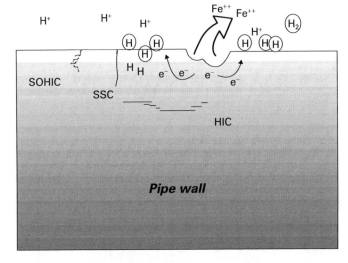

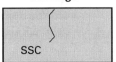

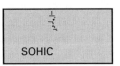

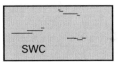

Figure 3.4   Schematic presentation of link between corrosion and various degradation mechanisms caused by hydrogen entry into steel

shown schematically in Figure 3.4. These phenomena may drastically reduce the integrity of the material, leading to leakages, brittle cracking and ruptures of the pipeline even at stresses well below the yield strength.

Hydrogen in the atomic stage is produced as part of the corrosion process, and in the presence of H_2S it exists for a sufficiently long time at the steel surface to become absorbed into the steel. Once inside the steel the hydrogen atom is free, unless trapped, to diffuse due to gradients in lattice dilation (i.e. at highly stressed/ strained zones), solubility (i.e. differences in microstructure), concentration (i.e. towards the outer surface) and temperature. If too much hydrogen is present at too high stresses in a susceptible microstructure the result will be hydrogen embrittlement cracking and loss of internal integrity.

However, the diffusing hydrogen may be trapped at non-coherent phase boundaries, as typically found in steels with a high density of planar inclusions and regions of banded structures, due to segregation of impurities and alloying elements. The trapped hydrogen atom will recombine to molecular gas, and be capable of exerting very high internal pressures. The high pressure may separate the internal surfaces, leading to local straining at the extremities of the phase boundaries and extension of planar cracking. Depending on steel chemistry, steel cleanliness, microstructure, and applied and residual stresses the net result may be one of following.

Sulphide stress cracking, SSC

This is a form of hydrogen stress cracking that involves embrittlement of the metal by atomic hydrogen. High strength steel and hard weld zones are particularly prone to SSC.

Hydrogen induced cracking, HIC

This consists of planar cracking resulting from pressurisation of trap sites by hydrogen. This is typically seen in steels with high impurity levels. Note that HIC may occur without externally applied stresses. When it occurs close to the surface it may result in blistering.

Stepwise cracking, SWC

This is cracking that connects hydrogen-induced cracks on adjacent planes in the steel wall. SWC is dependent upon local straining between the HIC, and embrittlement of the surrounding steels by dissolved hydrogen.

Stress oriented hydrogen induced cracking, SOHIC

This consists of staggered small cracks formed perpendicular to the principal stress (residual and applied) resulting in a 'ladder-like' crack array linking small pre-existing features akin to HIC. SOHIC is facilitated by high hydrogen concentration and local stresses at and above yield strength.

Soft zone cracking, SZC

This is a conglomerate of SSC and SOHIC that may occur in highly strained steels containing local 'soft zones', typically associated with highly strained welds in carbon steels.

Much effort is put into steel pipe manufacturing to ensure 'Sour Service Resistance'. This is obtained by keeping the hardness of base metal, heat-affected zones and weld metal at sufficiently low levels, and by improving steel cleanliness. Ultimately it may be necessary to use corrosion resistant alloys; see Sections 7.2.7 and 3.3.4 below.

3.3.4 *Steel microstructure and corrosion resistance*

In the linepipe industry the steels and the corrosion resistant alloys are sometimes characterised by their microstructure. In the following a brief description of steel microstructure is given, although for a pipeline engineer it is still mechanical properties and weldability that are the key issues.

Carbon steel microstructure

Linepipe steel is basically iron alloyed with small amounts of strengthening elements, primarily carbon and manganese. In molten condition and while above 750°C the iron atoms arrange themselves in a characteristic lattice, termed body-centred cubic. While in this state the steel phase is referred to as gamma phase or austenite. At lower temperatures the atom positions are changed to that of face-centred cubic lattice, resulting in a new phase called either alpha-phase or ferrite. Depending on the carbon content of the steel, some of the austenite is transformed into alternating layers of ferrite and cementite (a carbon rich constituent). This aggregate is termed pearlite (because of its mother of pearl-like visual appearance at high magnification), and the overall microstructure is referred to as ferrite-pearlite. Depending on how fast the temperature is lowered during the phase transformation from austenite to ferrite-pearlite it is possible to form other phases such as martensite and bainite. The latter can be further divided into upper and lower bainite. These phases causes internal stresses in the lattice, making the steel strong, hard and less ductile. Hardness and ductility may be controlled by a heat treatment called tempering. If the resulting grain size in the steel is too coarse (due to grain growth during hot forming while the steel is austenitic) it is possible to improve the situation by a normalising heat treatment, consisting of austenitising the steel, and repeating the phase transformation to ferrite-pearlite in a controlled manner.

Stainless steel microstructure

Stainless steel is characterised as steels with more than 12% chromium added as an alloying element. For low carbon steels, 12% chromium results in a martensitic microstructure. Increasing the chromium content to 17%, for example, allows the steel microstructure to be ferritic. If and when nickel is also added, the microstructure is changed to austenite. In general the corrosion resistance improves when going from martensitic through ferritic to austenitic stainless steels, at the expense of strength. Specially balanced alloys of iron, chromium and nickel use the best of each world, resulting in a duplex microstructure consisting of both austenite and ferrite, with good corrosion resistance and good mechanical properties. In austenitic and duplex stainless steel brittle phases, named sigma-, chi- and Laves phases, may form at grain- and phase boundaries during undue heat treatment or abusive welding conditions. In addition to being brittle, these phases also reduce corrosion resistance.

Nickel alloys

All alloys with nickel as the major element will be austenitic. In addition to the normal gamma phase, the atoms may compile in specially oriented arrays; gamma′ and gamma″. These structures improve both the mechanical properties and the corrosion resistance.

3.4 Sacrificial anodes

3.4.1 *General*

Sacrificial anodes are used for cathodic protection of the linepipe steel. See Section 6.7 for a description of the principles of cathodic protection.

For economic reasons the traditional sacrificial anode materials for marine application are alloys based on zinc or aluminium, although other metals may be used (such as carbon steel to protect stainless steel linepipe). Aluminium anodes are made electrochemically active by the introduction of small amounts of mercury or indium, the latter option being preferred for environmental reasons. To ensure a proper function of the anode the contents of other metals must be within certain limits, resulting in rigorous specifications of alloy composition.

Electrochemically the anode materials are characterised by the current capacity (measured in A h/kg) and the closed circuit potential (measured in V).

The only relevant physical parameter is the density (in kg/m^3).

3.4.2 *Zinc alloy*

Zinc alloy for sacrificial anodes is traditionally specified according to the US Military Specification MIL-A-18001 (1987), possibly modified as regards the content of aluminium or iron. A typical material composition is (cf. the Danish recommendations DS/R 464 and the DNV RP-B401) shown in Table 3.4.

The density of zinc is 7130 kg/m^3.

Theoretically, the electrochemical capacity of zinc is approximately 810 A h/kg, but in practice it is significantly lower. The closed circuit potential when the zinc anode is consumed in seawater is typically −1.05 V (with respect to an Ag/AgCl/ Seawater reference electrode), but higher (i.e. less negative) when the anode is buried in sediment.

Increasing the temperature of the anode results in a reduction of capacity and an increase of potential, i.e. a reduction of the driving potential difference to the steel. As mentioned in Section 6.7.4, zinc anodes should not be used at operating temperatures above 50°C. For operation between 35°C and 50°C the iron content

Table 3.4 Typical composition of zinc anode alloy

Element		Contents (%)
Aluminium	Al	0.10–0.50
Cadmium	Cd	0.125–0.015
Iron	Fe	Max 0.005
Copper	Cu	Max 0.005
Lead	Pb	Max 0.006
Silicon	Si	Max 0.125
Zinc	Zn	Remainder

Table 3.5   Typical composition of aluminium anode alloy

Element		Contents (%)
Zinc	Zn	2–6
Indium	In	0.005–0.05
Iron	Fe	Max 0.1
Copper	Cu	Max 0.006
Silicon	Si	Max 0.15
Others	Each	Max 0.02
Aluminium	Al	Remainder

should be lowered to a maximum of 0.002% and the aluminium content to a maximum of 0.2% to improve performance. This, on the other hand, tends to increase the propensity for cracking.

Values of electrochemical capacity and the driving potential recommended for practical design are given in Section 6.7.4.

3.4.3   *Aluminium alloy*

To improve the performance of aluminium a small amount of mercury or indium is added, where the latter tends to be preferred to reduce the environmental impact when the anode is consumed and dissolved in the seawater. A typical material composition is (cf. the Danish recommendations DS/R 464 and the DNV RP-B401) as shown in Table 3.5.

The NORSOK Standard M-503 is slightly more stringent with respect to the allowable impurities.

The density of aluminium is 2780 kg/m^3.

The electrochemical capacity of indium activated aluminium is typically 2600 A h/kg, decreasing approximately linearly with increasing temperature. The closed circuit potential of the anode in seawater is approximately −1.10 V (with respect to an Ag/AgCl/Seawater reference electrode).

Mercury activated anodes have a higher electrochemical capacity, but also higher corrosion potential, than indium alloy. Thus the driving voltage is lower, similar to that of zinc, and the corrosion is also less uniform.

Recommended values for practical design are given in Section 6.7.4.

3.5   Pipeline component materials

3.5.1   *General*

The components in a pipeline system other than the linepipe material itself are:

- valves;
- isolation couplings;

- flanges;
- branch connections (tees, wyes, olets, etc.);
- reinforcement sleeves (e.g. crack arrestors, buckle arrestors);
- brackets and support structures at valve stations and pipeline crossings.

In addition to relevant requirements according to their specific use, such components and fittings that are part of the pressure containment shall also meet the same requirements as the linepipe material.

3.5.2 *Component materials for sour service*

Pipeline components intended for sour service must have a fully documented resistance to sour environments. It is not enough to test a few samples of a lot of fittings, even if they belong to the same batch or heat of steel. For instance, if a lot of fittings belonging to the same batch or heat of steel have not been processed according to strict specification in order to maintain uniformity in sour service resistance, then one fitting will not necessarily represent all the pieces. Therefore, it is necessary to test all items individually.

3.6 Coating and insulation materials

3.6.1 *General*

Paints and coating materials are applied to the linepipe (and components) for purposes such as:

- internal drag reduction;
- external corrosion protection;
- thermal insulation;
- concrete weight coating;
- field joint coating.

Some material characteristics are given below. See Chapter 7 for a description of the coatings themselves.

3.6.2 *Material properties*

Materials used for the coating and insulation of pipelines are principally characterised by their density and/or thermal conductivity. Some typical values are given in Table 3.6, which for completeness also includes linepipe steel and seabed soil.

Table 3.6   Typical material parameters

Material	Density, ρ (kg/m³)	Thermal conductivity, k (W/m°C)
Linepipe steel	7850	45
Seabed soil	1500–2400	1.2–2.7
Concrete coating	1900–3800	1.5
High density polyoleofins	Variable	0.43
Fusion bonded epoxy	1500	0.3
Polychloroprene	1450	0.27
Solid polyolefins	900	0.12–0.22
Asphalt enamel	1300	0.16
Syntactic foams	Variable	0.1–0.2
Alumina silicate microspheres	Variable	0.1
Polyolefin foams	Variable	0.039–0.175

Chapter 4
Loads

4.1 Introduction

Loads on a marine pipeline, as on any structure, can be divided into the following categories:

- functional loads (actions resulting from the operation of the pipeline);
- environmental loads (normal actions from the natural environment);
- accidental loads (infrequent actions due to natural hazards or third party influence);
- installation loads (actions incurred during construction of the pipeline).

The distinction is not always obvious, and different national and international codes employ different classifications. Thus self-weight, buoyancy, hydrostatic pressure and soil reactions, which here are considered environmental loads, are classified as functional loads by DNV OS-F101. Actions from fishing activities are treated as accidental loads, but could also be considered functional or even environmental loads. Hydrostatic testing of the pipeline is an installation load, but could also be considered to be a functional load. In this chapter only the loads are identified, design procedures and mitigation measures are treated in Chapters 6 and 8, respectively.

4.2 Functional loads

4.2.1 *General*

Functional loads are defined as actions that result from the operation of the pipeline. Functional loads include the effects of:

- internal static pressure;
- pressure surge;
- operational temperature.

4.2.2 *Internal pressure*

The design pressure for a given location (reference height) is the maximum of the following two loads:

- maximum pressure during static conditions;
- maximum pressure during a combination of the static pressure and pressure surge.

Larger pressures can occur during pressure testing of the pipeline; see Section 4.5 on installation loads.

Pressure surge can occur in connection with the closure or opening of valves or a change in delivery pressure from compressors or pumps. Consequently, there is an intimate connection between the design pressure and the system for control of the pressure in the pipeline, as described in Section 4.2.3 below.

Traditionally, pressures are measured in bar, where 1 bar = $0.1 \text{ N/mm}^2 = 0.1$ MPa. In the absence of indications to the contrary, pressure values are taken to represent barg (gauge pressure), i.e. the pressure above the atmospheric pressure, which, by definition, is 1 bara (absolute pressure).

4.2.3 *Pressure control systems*

The pressure control system would normally comprise the pressure regulating system and the pressure safety system, as well as alarm systems and instrumentation to monitor the operation. The definitions are given in Figure 4.1.

The pressure regulating system ensures that the pressure in the pipeline is kept at or regulated to a certain level.

The pressure safety system denotes the equipment that independently of the pressure regulating system ensures that the accidental pressure in the pipeline is kept below set values, for example, it ensures that full well shut-in pressure cannot enter the pipeline. In this book the terminology maximum incidental pressure is adopted for the accidental load situation.

The above pressure definitions are not universally recognised. Thus pipeline design codes may identify the design pressure with the maximum operational pressure, and the notion of 'incidental pressure' may be embedded in the various safety factors. However, the DNV OS-F101 pressure definitions, reproduced in Figure 4.1, and their relation to the pressure safety system can be helpful, and a parallel also drawn when other design codes are used.

Pressure control example for gas systems

The pressure limiting systems protecting gas piping systems can be made to comply with the requirements stated in the Gas Piping Technology Committee (GPTC) *Guide for gas transmission and distribution piping systems*, 1998–2000.

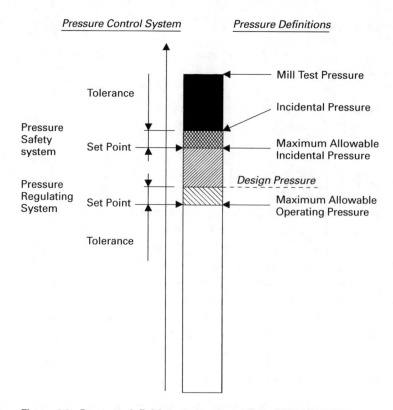

Figure 4.1 Pressure definitions (reproduced from DNV OS-F101)

Acronyms used in the context of design of the pressure control system include:

HIPPS   High Integrity Pressure Protection System
MAIP   Maximum Allowable Incidental Pressure
MAOP   Maximum Allowable Operating Pressure
ESD   Emergency Shut Down (valve)
PSD   Production Shut Down (valve)
PSV   Pressure Safety relief Valve
SCADA   Supervision, Control And Data Acquisition
SMTS   Specified Minimum Tensile Strength
SMYS   Specified Minimum Yield Strength

Pressure control systems consist of the following independent systems:

- a safety pressure limiting system to ensure that the MAOP of the piping and equipment of the land based terminal (compressor or receiving station) is not exceeded by more than 10%, as per the GPTC code;

- onshore equipment, consisting of thermal pressure safety relief valves for pressure protection of selected pipe sections, if temperature increases can result in pressures above MAOP + 10%;
- a safety pressure limiting system for protection of the offshore pipeline exiting the terminal to ensure that:
  □ the pressure in the onshore section does not exceed the MAOP + 10%, or the pressure that produces a hoop stress of 75% of SMYS, whichever is lower, as per the GPTC code. In this context the MAOP of the onshore systems is identical with the design pressure;
  □ the pressure in the offshore section does not exceed the design pressure of the submarine pipeline by more than 5% according to DNV OS-F101.

Note in the last point mentioned above, the split between the land based and the marine pressure limiting definitions. This follows the varying code pressure definitions as discussed in more detail in Section 6.2.5.

Overview and selection of pressure limiting systems

Designing a pipe system for the maximum accidental pressure without the need for pressure protection equipment is the simplest method for safeguarding a pipe system against overpressure. To evaluate the cost effectiveness of pressure protection the equipment costs can be compared with increased material costs. For a short, small diameter pipeline fed by a compressor any savings in pipe material and construction might well be offset by the life-cycle costs of overpressure protection equipment.

For longer, large diameter gas pipeline systems overpressure protection will, however, normally be designed. The level of protection required and the associated requirements to the protection system should be evaluated. Installation of safety pressure relief devices (pressure safety relief valves) is the basic system, and installation of a pressure limiting system which shuts off the gas supply to the pipeline is a more refined option. For gas pipelines ESD, or PSD for production flowlines, is normally mandatory, and the use of a HIPPS system is cost beneficial.

Pipelines where the well fluid pressure is driving the flow and pipelines intended for liquid service may also be designed with a HIPPS system, as a part of the overall field pressure control philosophy. Pipelines for liquids where rupture has a limited and not potentially fatal effect, for example, water injection lines and chemical/service lines, will naturally have less need for a pipeline safety system.

SCADA system

The operating pressure limiting system is part of the local Supervision, Control And Data Acquisition (SCADA) system, and forms part of the normal operation of the station. For a gas pipeline under normal operation the compressor station's discharge pressure will be controlled via a pressure-sensing device installed at the station's discharge valve. If a pre-selected MAOP is reached a switch will initiate an alarm in the SCADA system. If the pressure increases further a signal from a pressure-sensing switch will stop the operating compressor units by initiating the normal shutdown programme of the compressor units via the SCADA system. The set-point of the pressure switch will be such that the upper tolerance limit point is below the design pressure of the offshore pipeline.

ESD system for compressor station piping

The Emergency Shut Down (ESD) system includes a pressure-sensing device installed in the piping of each compressor train upstream of the discharge valve. In the case where the SCADA alarm system is insufficient to stop a compressor unit, the signal from the pressure switch will initiate a mechanical shutdown of the respective compressor train via its ESD system. The set point of the shutdown pressure will be such that the upper tolerance limit point will be below the MAOP of the compressor station piping.

Furthermore, the set point will have to be selected to be sufficiently low for the ESD system to protect the downstream pipeline against excessive incidental pressures. Restarting the compressor unit will only be possible after manual reset of the ESD system. The ESD system is separated from the SCADA system of the compressor station in such a way that signals for ESD actions are separated from normal process signals. The ESD data are transmitted to the SCADA system.

Safety pressure limiting system for the offshore pipeline (HIPPS)

The High Integrity Pressure Protection System (HIPPS) works independently of the SCADA and ESD systems on the compressor station. The pressure-sensing elements of the HIPPS system are installed in the piping of each compressor train upstream of the discharge valve. In the case where the ESD system fails to stop a compressor unit and the pressure continues to rise, the HIPPS system will interrupt the turbine fuel or electric power supply. At the same time a signal from the HIPPS system will initiate the emergency shutdown of the compressor unit and its utilities.

The set point of the shutdown level must be chosen so that the upper tolerance limit point will be below the pressure level that is governing for the onshore section of the pipeline, i.e. a MAOP + 10% or a hoop stress of 75% of SMYS. Furthermore, the limit point shall be below the design pressure + 5% of the submarine section of the pipeline. Restarting the compressor unit will only be possible after manual resets of the HIPPS system and of the ESD system.

The reliability of the HIPPS system should be documented to be ten times greater than that of an equivalent relief system. A reliability study should be performed, and the necessary redundancy determined. The HIPPS system should always be applied in combination with a pressure alarm and be failsafe.

The HIPPS system should meet the requirements of international standards like the DIN V 19250/19251 *Control technology* or the IEC 61508 *Functional safety of electrical/electronic/programmable electronic safety-related systems.*

4.2.4 *Temperature*

Temperature differences will often occur between the transported medium and the surroundings. The effects of such temperature differences shall be taken into account in connection with the design of the pipeline system. This includes low temperatures in gas pipelines as a result of the cooling associated with pressure drop (Joule–Thompson effect).

More importantly, any difference between the operating temperature and the temperature when the pipeline was installed on the seabed will give rise to deformations and/or stresses. Temperature is a major contributor to pipeline loading, and the temperature in the pipeline will vary with the flow conditions. Production flowlines will suffer shutdowns, and with a time lag they will cool and subsequently partially recover end expansions and/or lateral buckling. This introduces cyclic loads that again can introduce low cycle fatigue or ratcheting effects; see Section 6.5.

Thus the actual temperature profile along the pipeline must be considered in connection with the design of the pipeline; see Section 2.3.4.

4.3 Environmental loads

4.3.1 *General*

Environmental loads are defined as actions resulting from the interaction of the pipeline with its environment. In addition to gravity forces (self-weight, buoyancy and hydrostatic pressure) environmental loads are primarily generated by wave and current action. Other loads that can be characterised as environmental are soil pressure and other natural actions, including the temperature of the surroundings. A number of environmental loads are either self-explanatory or may be very site or project specific. The following sections therefore deal exclusively with loads generated by waves and current.

4.3.2 *Hydrodynamic forces*

A pipeline near the seabed is exposed to hydrodynamic forces from wave and current action. The force variation is in general quite complex, and simple analytical expressions can only describe the force variation in an approximate manner. The following paragraphs present a brief discussion of the hydrodynamic forces on a pipeline exposed to different flow conditions, and introduces practical expressions for calculating the forces for application in pipeline design.

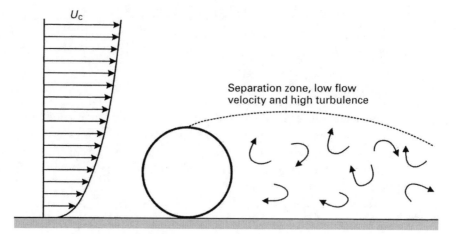

Figure 4.2   Pipeline exposed to steady current

Steady current

In general only the flow component perpendicular to the pipeline is considered. A pipeline near or on the seabed represents an obstruction for the flow. The hydrodynamic force is generated by the combined effect of increased flow velocity above the pipe and the flow separation from the pipe surface at the leeward side; see Figure 4.2.

The hydrodynamic force can be expressed by two force components, one in line with the flow (drag force) the other perpendicular to the flow (lift force). The two force components can be calculated by analytical expressions:

In-line:

$$F_D = \frac{1}{2} \rho D C_D U^2$$

Lift:

$$F_L = \frac{1}{2} \rho D C_L U^2$$

where

F_D	drag force
F_L	lift force
ρ	density of seawater
U	undisturbed current velocity perpendicular to the pipeline averaged over the height of the pipeline
C_D	drag coefficient
C_L	lift coefficient.

The drag and lift coefficients have been determined through a number of experimental investigations (e.g. Jones 1971), and depend primarily on the pipe surface roughness described by the roughness ratio k/D and the Reynolds number $Re = UD/\nu$.

Wave flow or combined wave and steady current flow

The pressure gradients accelerating the flow under wave action induce a horizontal force on the pipeline, called the inertia force. Only the pressure gradient perpendicular to the pipeline is considered, and the force can be assessed using potential flow theory:

Inertia force:

$$F_M = \frac{\pi}{4} \rho D^2 C_M a$$

where

ρ	density of seawater
a	wave induced water particle acceleration, $a(t)$
C_M	inertia coefficient
t	time.

The inertia force may be considered as the sum of two components:

$$F_M = \frac{\pi}{4} \rho D^2 a + \frac{\pi}{4} \rho D^2 (C_M - 1) a$$

The first component describes the force from the pressure gradient on the body of water replaced by the pipe (the Froude–Krylov term), the second reflects the influence of the pipeline on the water motion (the added mass term).

Near the seabed the pressure gradient and water acceleration follow the general seabed contour, thus the acceleration term does not contribute to the vertical force on the pipeline.

A practical approach to calculating the total in-line force from combined wave and current flow is simply to add the inertia and the drag components. The drag component is calculated using the instantaneous total water velocity perpendicular to the pipeline. The result is called the Morison equation:

In-line:

$$F_H = F_D + F_M = \frac{1}{2} \rho D C_D U |U| + \frac{\pi}{4} \rho D^2 C_M a$$

The lift force is not dependent on the acceleration, and an expression similar to that used for calculating the drag force can be used, see Figure 4.3.

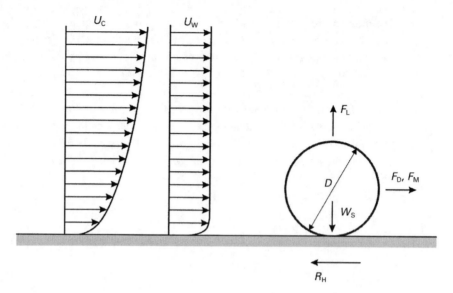

Figure 4.3   Forces on a submarine pipeline exposed to wave and current action

Lift:

$$F_{L} = \frac{1}{2} \rho D C_{L} U^2,$$

where

F_D	drag force
F_M	inertia force
F_H	in-line force
F_L	lift force
ρ	density of seawater
U	water particle velocity, i.e. sum of wave and current induced velocity $U_w(t) + U_c$
a	wave induced water particle acceleration, $a(t)$
C_D	drag coefficient
C_M	inertia coefficient
C_L	lift coefficient
t	time.

The above expressions have been used extensively and with success in pipeline design. It is, however, important to understand that the expressions give only a very simplified reproduction of the hydrodynamic forces experienced by a pipeline exposed to a natural irregular sea state, possibly under the presence of a steady current. See Sarpkaya and Rajabi (1980), Grace and Zee (1981) and Bryndum *et al.* (1983).

The unsteady and reversing flow induced by waves gives rise to very complex structure–flow interactions, and the resulting hydrodynamic forces cannot be fully reflected in the two equations and three coefficients. During the passage of the wave crest a wake is developed at the leeward side of the pipeline. When the free stream flow starts to reverse (at the passing of the wave trough) the wake is washed over the pipeline, resulting in high local flow velocities, and the pipeline will experience high lift and drag forces out of phase with the general undisturbed free stream velocity. This is one aspect of the water–structure interaction that cannot be described by the simple force expressions. Other aspects are:

- the development and variation of the boundary layer and the separation zone on the pipeline;
- the formation and development of the upstream separation zone;
- the time varying seabed boundary layer.

Jacobsen *et al.* (1984) have demonstrated that the local near-pipe flow velocity can be used to calculate the time variation of hydrodynamic forces with high accuracy. Lambrakos *et al.* (1987) and Verley *et al.* (1987) have made the same observation, and developed a numerical model (wake model), capable of accurately predicting wave-induced forces.

A number of research projects have been undertaken in order to understand the hydrodynamic forces on pipelines, and with the purpose of developing rational methodologies for on-bottom stability design. The two most extensive are PIPESTAB (Wolfram *et al.* 1987) and the On-bottom Stability Project by PRCI (Ayers *et al.* 1989).

4.3.3 *Hydrodynamic force coefficients*

Although unable to predict the time variation accurately, the simple expressions for calculating hydrodynamic forces are used in practical design. The main reason for this is that it is possible to select force coefficients that, in combination with the simple expressions, will result in reasonable requirements to on-bottom stability design, see Chapter 6. The force coefficients C_D, C_M and C_L cannot be found using analytical methods only, model tests or complex numerical flow simulations are required, and some typical values are presented below.

The three force coefficients, C_D, C_M and C_L, depend on a number of parameters, for example, the relative pipe roughness (k/D), the relative amplitude of water motion (or the Keulegan–Carpenter number, KC), and the ratio between the steady current and the wave velocity. The force coefficients C_D and C_L are presented in Figures 4.4 to 4.7. The coefficients presented give the best overall fit between forces measured experimentally and forces calculated using the above equations. Practical experience has shown that the use of the theoretical value $C_M = 3.29$ for the inertia coefficient results in adequate accuracy in the

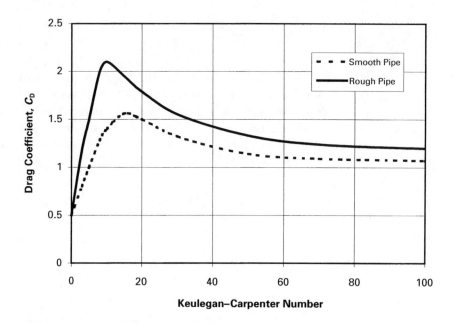

Figure 4.4   Drag coefficient against Keulegan–Carpenter number – pure wave flow

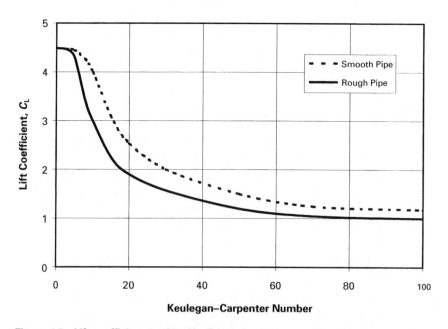

Figure 4.5   Lift coefficient against Keulegan–Carpenter number – pure wave flow

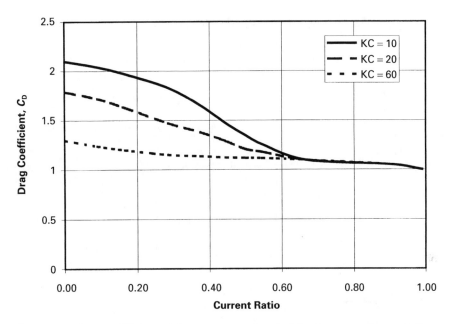

Figure 4.6   Drag coefficient against current ratio, combined wave and current flow. Values for rough pipe are presented ($k/D \approx 10^{-2}$)

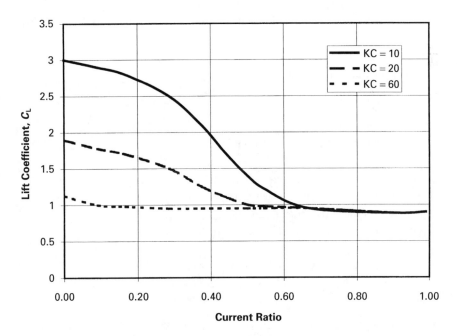

Figure 4.7   Lift coefficient against current ratio, combined wave and current flow. Values for rough pipe are presented ($k/D \approx 10^{-2}$)

force calculations. Therefore, it is recommended that $C_M = 3.29$ be used for all pipelines resting on the seabed.

The force coefficients have been found experimentally as presented in September 1993 by the American Gas Association (AGA, now PRCI) *Submarine pipeline on-bottom stability, Volume 1, Analysis and design guidelines*, see also Bryndum *et al.* (1992).

The force coefficients are plotted as functions of:

KC	Keulegan–Carpenter number $(U_w T/D)$
T	wave period
α	current to wave ratio $(U_c/(U_c + U_w))$
k/D	non-dimensional pipe roughness.

Smooth pipe corresponds to a roughness $k/D \approx 10^{-3}$, rough pipe to $k/D \approx 10^{-2}$.

4.4 Accidental loads

4.4.1 General

Accidental loads are defined as loads which have a low probability of occurrence. For marine pipelines such loads may be grouped into the following.

Natural hazards, for example:

- earthquakes;
- mudslides;
- iceberg scouring.

Third party hazards, for example:

- dropped objects (near platforms);
- fishing activities (trawling);
- shipping (anchoring, grounding, sinking);
- military activities (firing, unexploded ordinance).

To the extent that it is not possible to protect the pipeline totally against accidental loads, a risk analysis must be performed; see Chapter 5.

4.4.2 Dropped object loads

The main source of dropped objects on pipelines is material handling at or on drilling rigs or well intervention vessels. Drop energies and hit frequencies are found from analysis of the topical equipment handling operations. As an example,

Table 4.1 Typical data for dropped objects

Object	Weight (t)	Lift frequency (per well operation)	Impact energy (kJ)	Flowline hit frequency (per well operation)
BOP stack	220	4	20 732	2.0×10^{-15}
Coiled tubing reel	45.0	1	2507	1.9×10^{-10}
Running tool	13.0	1	1816	2.0×10^{-9}
Skid	16.2	3	971	2.9×10^{-10}
X-mas tree	24.4	1	393	2.5×10^{-10}
9 1/2″ drill collar	3.0	1	228	6.8×10^{-9}
30″ casing	5.5	15	176	1.0×10^{-7}
Container/basket	8.2	~780	133	3.5×10^{-6}
20″ casing	2.33	26	48	2.7×10^{-7}
Casing/tubing/pipe	1.25	~160	<15	3.3×10^{-6}
Other			~<40	$~4.0 \times 10^{-7}$
Total hit frequency				7.3×10^{-6}

Table 4.1 reproduces typical figures for such a study for a North Sea field (subsea installation).

4.4.3 Trawl loads

The dominant hazard to pipelines from fishing activities is interaction with bottom trawls, including impact, trawl pull-over and (in special cases) hooking. Design loads may be determined in accordance with the DNV Guideline No 13 *Interference between trawl gear and pipelines*. Bottom trawl gear is commonly divided into otter trawls, where the trawl bag is kept open by individual trawl boards, and beam trawls, which use a transversal beam with trawl shoes at the ends.

The total mass of a beam trawl may be up to 5500 kg, and a typical beam trawl configuration is shown in Figure 4.8.

Otter trawl boards up to 5000 kg are used, but a common polyvalent trawl board will typically have a maximum steel mass of 4000 kg. With a trawling velocity of 5 knots (2.57 m/s) and considering the added mass the trawl board impact load is 16.5 kJ. However, studies by DNV indicate that the actual energy on the pipeline is of the order of 7.5 kJ. Some trawlers use two nets in a twin configuration, somewhat like the beam trawls of Figure 4.8, but the nets are close together, and are held open by two outer trawl boards and a central clump weight, which is a cylindrical or spherical concrete block that rolls on the seabed. The mass of the clump weight is likely to be higher, but the round form makes the impact less critical.

The pull-over force depends upon the trawling velocity, the trawl mass, and the stiffness and length of the towing wire (warp). DNV Guideline No 13 gives the formula:

$$F_p = C_F v (m_t EA_w / L_w)^{1/2}$$

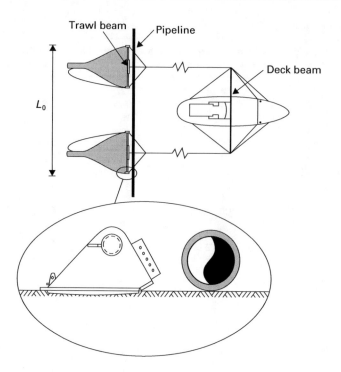

Figure 4.8   Typical beam trawl (top view and trawl shoe detail)

where

F_p pull-over force
v velocity
m_t trawl mass, including added mass
EA_w warp stiffness
L_w warp length
C_F empirical constant.

For beam trawls, the constant C_F ranges from 3 to 5, depending upon trawl geometry, particularly whether or not a hoop bar is mounted in front of the trawl shoe.

For trawl boards the constant is strongly dependent upon the level of the pipeline relative to the seabed, and for polyvalent and rectangular boards the DNV Guideline No 13 offers the expression:

$$C_F = 6.6(1 - \exp[-0.8(D + 2H_{sp} + 0.4)/H_B])$$

where

D outer diameter of (coated) pipe (in m)

H_{sp} height of the span (in m)

H_B height of the trawl board (in m).

The span height H_{sp} is taken to be negative for pipelines that are partly buried.

4.5 Installation loads

Installation of marine pipelines is to a great extent weather dependent, and part of the installation engineering is the determination of the acceptable limits (wind speed, wave height, current) for the installation to take place.

Actions during the construction phase should be carefully evaluated for the specific method of construction, i.e.:

- transport of pipe joints (bare and coated);
- installation of pipe strings (laying, reeling, towing, pulling);
- tie-in;
- trenching and backfilling;
- hydrostatic testing.

Apart from the pipeline self weight and the normal environmental loads, the specific actions during installation will mostly be imposed static and dynamic deformations (from laybarge stingers or reel barge spools, tie-in tools or davit lift operations, trenching equipment, etc.). An exception is hydrostatic testing, where the test pressure is normally prescribed by regulations, typically corresponding to 15% above the design pressure, although substantially different values may be specified.

Chapter 5
Risk and safety

5.1 Introduction

The overall safety concern for a marine pipeline is to ensure that during both construction and operation of the system there is a low probability of damage to the pipeline, or to detrimental impact on third parties, including the environment. The risk and safety activities discussed below are related to the system safety during operation only, and do not include the occupational health and safety for the personnel during the construction activities.

Consequently, risk and safety activities in relation to offshore pipeline projects have the following main objectives:

- security of supply;
- personnel safety;
- environmental safety.

The specific focus on any of the above mentioned items would depend on the medium to be transported in the pipeline system. For transport of natural gas the environmental impact may be less severe compared with systems transporting oil, but the safety of personnel may be more critical due to the potentially explosive release of the gas medium.

This chapter includes a description of the methodology known as risk management that is implemented to achieve the overall goal for the risk and safety activities. The following items have been included in the description:

- safety policy;
- risk management;
- risk acceptance criteria;
- risk assessment.

The risk and safety activities can be performed at different stages in the project development process. In the initial phase an overall risk assessment (security of supply) can be performed to evaluate whether a single pipe solution is sufficient, or whether redundancy is needed in the form of dual offshore pipelines or another type of back-up system (e.g. storage facilities).

Even though an offshore marine pipeline in general must be considered to be a very safe structure, there are special hazards that must be considered in connection with the design and construction. The following special risk aspects are discussed:

- sub-sea isolation valves;
- welded or flanged connections;
- corrosion in high pressure/high temperature (HP/HT) pipelines.

5.2   Safety policy and philosophy

In the initial phase of a project it is important to establish the safety policy and philosophy. The safety policy should clearly state the main safety objectives for the project. These might, for example, be that:

- The pipeline shall have the same safety as other pipelines in the area.
- A severe damage of the pipeline may occur no more frequently than every 100 years of operation.

In accordance with the overall safety policy and philosophy the marine pipeline should be designed, constructed and maintained in an economic way, and should be fit for use during the design working life. In particular, the pipeline should fulfil the following requirements:

- perform adequately under all expected actions;
- withstand extreme and/or frequently repeated actions occurring during construction and anticipated use.

The requirements are often described through a set of limit states in accordance with normal design practice. A limit state is defined as the threshold between states where the structure functions according to the prescribed conditions, and states where this is not the case.

When the above safety format is adopted, the limit states are normally divided into Ultimate Limit States (ULS), which concern the maximum load carrying capacity, and Serviceability Limit States (SLS), which concern the normal use. It has also been customary to introduce Accidental Limit States (ALS), concerning rare and extreme conditions. Recent practice, for example that adopted by the Eurocode system, is to treat these as ULS cases, sometimes referred to as Accidental Load Situations.

A limit state may be irreversible or reversible. These are defined as follows.

Irreversible

The structure remains at the limit state when the actions which caused the condition are removed. Examples are:

- the collapse of a structure or part of it;
- permanent unacceptable local damage;
- permanent unacceptable deformations.

Reversible

The structure will not remain at the limit state when the actions that caused the condition are removed. Examples are:

- temporary expansions;
- temporary large deflections.

In general ULS are irreversible, whereas SLS can be both irreversible and reversible.

5.3 Risk management

5.3.1 *General*

Systematic risk management is an important part of the planning of a pipeline project. The risk management will provide input to route selection and the protection strategy for a pipeline. Risk management for pipeline projects follows the same procedure as for all other types of infrastructure projects.

The overall methodology for risk management is illustrated in Figure 5.1. The level and detail of the risk management process will depend upon the actual conditions for the pipeline project. In a well-known area risk assessment may be of less importance than for areas where there is less experience with installation and operation of marine pipelines.

As seen in Figure 5.1, the risk management process comprises a number of steps, which are detailed and further described in the following sections. It is important that the risk management process is included as an integrated part of the entire design and construction activities for a pipeline project. The required protection strategy should be established in the design phase, so the recommended risk reducing measures can be included in the project. Guidance on integrity management of gas pipelines may be found in ASME B31.8S *Managing system integrity of gas pipelines*, which is a recent supplement to ASME B31.8.

The risk is conventionally quantified as the product of the frequency and the consequence of a given event. The first step in the risk analysis is to

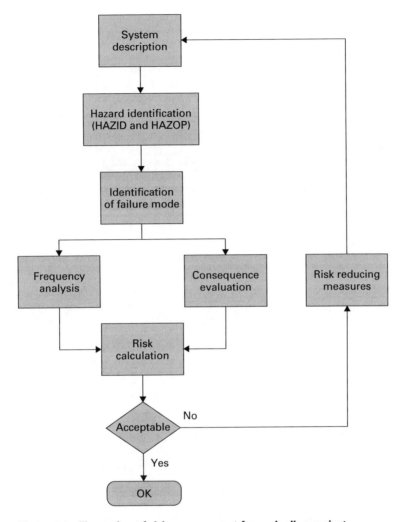

Figure 5.1   Illustration of risk management for a pipeline project

identify possible hazard scenarios, such as sinking ships, seismic events, sabotage, etc.

A clear organisation of the risk and safety activities is of vital importance. Risk management will only have the desired effect if supported by the top management of the project. The project manager should have overall responsibility for the risk management system, and be responsible for approval of all hazard identification, which is the starting point for the risk management.

For the installation phase, guidance may be found in the recent recommended practice DNV RP H101 *Risk Management in Marine and Subsea Operations* issued by Det Norske Veritas. The recommendation is based upon the ALARP principle. See Section 5.4.1.

5.3.2 *Hazard identification*

Hazard identification is a process whereby all possible hazards that can threaten the integrity of the pipeline are identified through brainstorming sessions. The brainstorming sessions are often denoted HAZID and HAZOP workshops. Personnel with detailed knowledge of the pipeline system and the process conditions should participate in the workshops, which should be headed by a specialist in risk and safety and with expertise of conducting HAZID and HAZOP workshops.

Before the workshop starts, a review of 'accident history' for other pipelines/cables in the project area should be performed. As accidents are rare there may be no direct location specific data to form the basis for the hazard identification. Therefore, accident history for other locations will also be relevant as start-up information for hazard identification.

The results of the workshops should be documented in a systematic way. The documentation can be established as a database, where all identified hazards are recorded on a so-called hazard sheet. The hazard sheet contains information on the type of hazard, as well as information about the cause of the hazard.

Furthermore, the hazard sheet should include information on the frequency and consequence levels that have been established during the risk assessment. An example of such a hazard identification sheet is shown in Figure 5.2.

After the HAZID workshops have been finalised, a review of the identified hazards should be performed. The review should focus on the most critical hazards identified, and a decision should be taken on how to analyse the identified hazards.

A method often used for analysing the operational hazards is HAZOP analysis. The HAZOP analysis includes the following activities:

- the break-down of the system into pipe segments and main plant items;
- a review of flow sheets and P&IDs, etc.;
- application of guide-words to different process parameters to identify possible deviations, using standardised guide-words to identify disturbances;
- a description of potential consequences of the disturbances;
- identification of potential causes of disturbances in cases with significant consequences.

Table 5.1 gives an overview of guide-words. The guide-words are applied to process parameters such as flow, pressure and temperature, and to any operations of the system such as testing, start-up, shutdown and maintenance.

Other methods can be used to analyse the identified hazards, such as a fully quantified risk analysis (QRA), where frequency and consequences are established for each hazard.

Identification

Hazard no.	3
Revision no.	1
Revision date	2001-05-09

Categorisation

Project phase	Operation
Components	Line pipe Valves at midline tie-in Pipe coating

Hazard Description

The pipeline is damage by a dropped or dragged anchor.

Cause and Comments

An anchor is dropped from a ship either as a controlled drop in connection with an emergency anchoring operation or the anchor is dropped by accident. The pipeline is subsequently damaged by the dragging anchor.

Risk Assessment

Consequence	3
Frequency	3
Risk level	ALARP

Risk Types

Delays/Cost

Doc. Reference

31.02.0009

Risk Reducing

Implemented	Planned		Possible
Pipeline route chosen to minimise the exposure to heavy ship traffic. A risk assessment has been performed. See ref. 31.02.0009.	Pipeline to be trenched in accordance with the results in ref. 31.02.0009. Plans for repair of the pipeline in the soft soil area shall be elaborated.	2001-04-20	The pipeline may be trenched into the seabed to lower the risk. Lay-out of protection structure for the mid-line tie-in.

Figure 5.2   Example of hazard sheet for a pipeline project

5.4   Risk acceptance criteria

5.4.1   *General*

Operational risk acceptance criteria should be based on the overall safety policy and philosophy for a given project. Such criteria can follow general company policy, be based on internationally recognised standards, or be developed especially for a given project. A recognised standard may be the recommended practice DNV RP F107 *Risk assessment of pipeline protection* issued by Det Norske Veritas.

The mean individual risk may be expressed by the Fatal Accident Rate (FAR), which is defined as the number of fatalities by accident per 10^8 personnel hours. For the operating phase a typical criterion would be FAR < 10, and for the

Table 5.1   Overview of guide-words for HAZOP analysis

Guide-word	Meaning	Example
No/not	The complete negation of the intentions	No flow when production expects it
More	Quantitative increase	Temperature higher than assumed in design
Less	Quantitative decrease	Lower pressure than normal
As well as	Quantitative increase	Other valves closed at the same time (logic fault or human error)
Part of	Quantitative decrease	Only part of the system is shut down
Reverse	The logical opposite of the intention	Back-flow when the system shuts down
Other than	Complete substitution	Liquids in the gas piping

installation phases similar criteria will be established. Personnel exposure includes shuttling to and from the facility in question.

A typical acceptable failure probability for a single flowline or interfield pipeline is 10^{-4} per year, which is the figure specified by DNV OS-F101 for normal safety class. For a long trunkline or an entire pipeline system the failure probability is more reasonably expressed as events per kilometre per year, and the pipeline owner should establish a safety policy, including specific risk acceptance criteria, which should be approved by the relevant authorities.

Often the risk acceptance criteria operate within a domain between the clearly unacceptable and the clearly acceptable, where the risk shall be As Low As Reasonably Practicable (ALARP), which means that risk should be reduced as far as technically and economically feasible. The ALARP principle is illustrated in Figure 5.3.

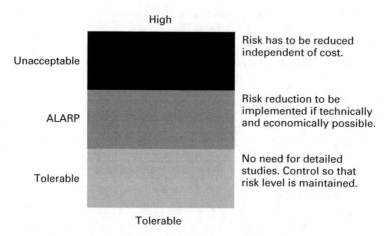

Figure 5.3   Illustration of the ALARP principle

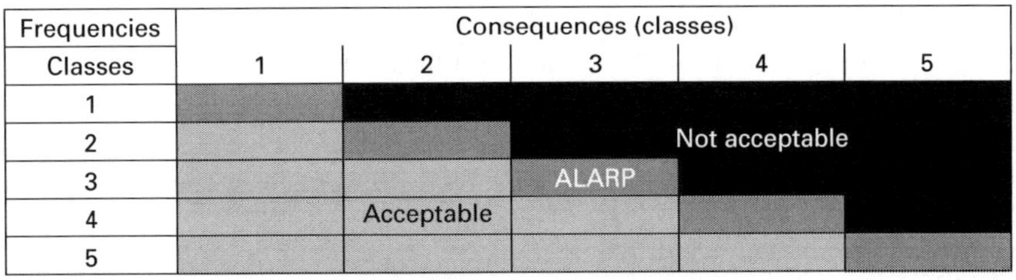

Frequencies	Consequences (classes)				
Classes	1	2	3	4	5
1				Not acceptable	
2				Not acceptable	
3			ALARP		
4		Acceptable			
5					

Figure 5.4   Example of a risk matrix

The ALARP principle implies that some low cost risk reducing measures should be introduced even if the risk is considered to be acceptable. Risk acceptance criteria can be developed in terms of a so-called risk matrix, which is a relationship between the frequency and the consequences of unwanted events. Figure 5.4 shows an example of a risk matrix.

The consequence classes can be defined in terms of human fatalities, loss of property and environmental impact. The frequency classes are defined based on the actual project as established in the safety policy for the project.

5.4.2   *Cost benefit analysis*

A cost benefit analysis is an appropriate method to evaluate the gained risk reduction versus the extra cost. The cost benefit value (CBV) is defined as

$$CBV = \frac{\Delta Cost}{\Delta Risk}$$

where

$\Delta Cost$	increased cost due additional risk reducing measures
$\Delta Risk$	reduced risk due to the measures.

A cost-effective solution will give a ratio of less than one, whereas a non cost-effective solution will give a ratio greater than one. For practical computations the CBV can be calculated as:

$$CBV = \frac{C_{RM}}{\sum_{n} \frac{\Delta C_R + \Delta C_P}{(1 + r)^n} \cdot P_f}$$

where

C_{RM}	cost of the risk reducing measure
ΔC_R	reduction in repair cost

ΔC_p reduction in production loss

P_f probability of failure or the failure frequency if the frequency is small

r interest rate

n number of years considered.

All cost measures are real and adjusted for inflation.

From the formula it is seen that the reduction in repair cost and production loss must be significantly greater than the cost of the risk reducing measures when the failure probability is low (10^{-3}–10^{-4}).

Whichever of the above mentioned acceptance criteria is selected for a given project, it is important that an ongoing follow-up on the acceptable risk is performed. This can be done in hazard sheets, with reference to the risk assessments performed.

The risk acceptance criteria depend on several factors, including the purpose of the pipeline, and in general it is up to society to determine the acceptable risk level. This is seen from the point of view of environmental impact, security of supply, loss of property, etc. The pipeline owner may have his own in-house risk acceptance criteria for pipeline systems.

5.5 Risk assessment

5.5.1 *General*

As part of the overall route planning of a pipeline, a risk analysis should be performed to justify that the chosen pipeline route and pipeline protection strategy will result in an acceptably low probability for severe damage or rupture.

The methodology for the risk analysis is identical for each hazard, namely that the frequency and consequences of hazards are established. The risk is then defined as the product of the frequency and the consequences:

$$Risk = f \times C$$

In principle, the accident frequency is established on the basis of statistical information or appropriate mathematical modelling of a situation that can result in damage of the pipeline. The consequences can be defined in various ways, depending on the risk policy that is adopted. Typical consequences are system downtime, loss of property, fatalities and pollution of the environment.

The most critical hazards will depend on the specific project in question. For pipeline projects connecting platforms, the risk related to near platform areas and the risers is often of great importance. Examples of near platform hazards are dropped objects, ship collision against risers, interference with approaching drilling rigs, etc. Such hazards should be analysed, and the required protection strategy established.

A convenient method of reducing the risk in connection with dropped objects is to route the pipeline approach to avoid areas with a high potential for dropped objects, typically areas defined by the operating ranges of the platform cranes.

For pipelines passing areas with high ship traffic the most critical hazards may be related to dropped and dragged anchors, sinking ships, grounding ships, etc. Risk analyses of such hazards should be based on statistics from the specific area, and combined with general statistics for ship accidents. See Section 5.5.3.

The consequence modelling of an accident scenario will depend on the way the acceptance criteria have been established. If the acceptance criteria are based on damage/rupture of the pipeline, only structural evaluations of the pipeline are required. However, such structural evaluations are associated with uncertainties, and often a number of assumptions have to be made. The evaluation of structural consequences may be performed by rough models based on engineering judgement or by detailed computer modelling.

If the acceptance criteria are based on the consequences for people and/or the environment it is necessary to perform consequence modelling of the release of gas, oil, etc.

5.5.2 *Risk reducing measures*

An important part of risk management is the introduction and evaluation of risk reducing measures. Such risk reducing measures depend on the hazards against which the system is to be protected.

Typical risk reducing measures are:

- trenching of pipelines to a specified depth;
- rock cover on pipelines;
- protective covers on valve assemblies, tees and spools;
- protection of risers against dropped objects and ship collision.

Protection of risers against ship impact is normally achieved by the installation of riser guards, i.e. tubular steel structures, that are welded or bolted to the platform.

According to the overall methodology for the risk management, risk reducing measures should be considered in cases where the risk is found to be unacceptably high or the risk is in the ALARP domain. If the risk is in the ALARP domain the decision on introducing risk reducing measures is based on a cost benefit analysis.

5.5.3 *Example: Risk of anchor damage*

The application of risk assessment to determine the required trenching depth for a pipeline crossing a heavily trafficked shipping lane is presented as an illustrative example. The overall methodology is illustrated in Figure 5.5. See the general risk

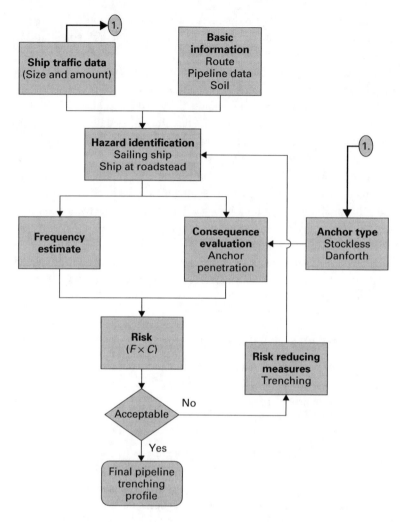

Figure 5.5 Overview of methodology in the risk assessment process

management process described in Section 5.3.1. Note that the ship traffic data (1.) provide input not only to the hazard identification, but also to the anchor type design.

Figure 5.6 shows the assumed ship traffic distribution along a pipeline crossing a shipping lane.

The distribution is based on the following factors:

- the numbers and sizes of ships passing the pipeline;
- the yearly probability for dropping an anchor;
- the anchor dragging length;
- anchor penetration;
- soil conditions.

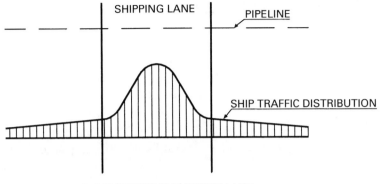

SHIPPING LANE

PIPELINE

SHIP TRAFFIC DISTRIBUTION

PIPELINE CROSSING SHIPPING LANE

Figure 5.6   Traffic distribution at shipping lane

To determine the required trenching depth, both the frequency and consequences of an anchor crossing must be determined and reassessed against appropriate risk acceptance criteria. The analysis follows the following steps: (1) the number of anchor crossings is determined, and (2) the consequences are established based on a geometrical consideration of the pipeline–anchor interaction.

The probability P_{drop} for an anchor interfering with the pipeline can be determined by the following formula:

$$P_{drop} = f_{drop} \sum_{}^{ship\ groups} n L_{crit} \frac{1}{V_{ship}} \frac{1}{1852} \text{ (penetrating anchors/year)}$$

where

n	number of critical ship movements per year
f_{drop}	frequency of drops per year
V_{ship}	speed of ship (knots)
L_{crit}	dragging length for dropped anchor (m).

An anchor can be dropped either in an emergency situation (risk of ship grounding) or by accident. Both scenarios have been observed in reality.

The crossing frequency can be calculated as 10^{-3}, based on the following assumptions:

- 50 000 ships passing per year;
- a drop frequency of 10^{-6};
- an anchor dragging length of 100 m;
- a constant ship velocity of 15 knots.

The annual damage rate from anchors, P_f, can be calculated by the following formula:

$$P_f = P_{drop} \times P_{damage}$$

where

P_f	annual failure probability for damage (in a design situation this frequency will be replaced by the risk acceptance criterion)
P_{drop}	annual probability of a dropped or dragged anchor across the pipeline
P_{damage}	conditional probability for damage given an anchor crossing

As mentioned above, it is necessary to establish acceptance criteria for the pipeline to determine the required burial depth. As described in Section 5.4, such acceptance criteria can be established in a number of ways.

Based on an acceptance criterion of 10^{-4} per year per kilometre for pipeline damage due to anchor impact, the above formula can be rewritten to determine the conditional probability for severe damage of the pipeline given that an anchor crosses the pipeline as follows:

$$P_{damage} = P_f/P_{drop} = 0.10$$

In this formula the annual damage probability has been replaced by the acceptance criterion.

The conditional probability that an anchor dragged across the pipeline area will damage the pipeline depends on the anchor penetration and the soil cover on the pipeline. Therefore, to determine the conditional probability of an anchor hitting the pipeline it is necessary to establish the relationship between the ship size and the anchor penetration.

An anchor and a pipeline are illustrated in Figure 5.7. The consequences for the pipeline in the case of an impact will depend on the position at which the anchor hits the pipeline.

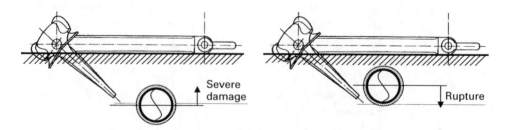

Figure 5.7 Anchor and pipeline impact

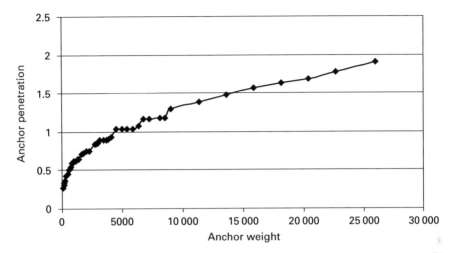

Figure 5.8   Anchor penetration (in m) vs anchor weight (in kg)

For relatively hard seabed material the typical anchor penetration will be in the order of one half to one fluke length. For soft soil the anchor penetration may be significantly higher (several metres), and in soft material it may not be feasible to protect the pipeline against dropped and dragged anchors. Figure 5.8 gives a relationship between the anchor penetration and the anchor weight, based on the assumption that the anchor penetration will be 0.7 times the fluke length.

The weight of an anchor will depend on the size of the ship, which again means that the anchor penetration will be a function of the ship size. For simplicity in this example the ship traffic is distributed into three classes according to Table 5.2. Each ship class is presented by an average dead weight tonnage (DWT), and for each ship class the anchor weight is estimated. Based on the estimated anchor weight the penetration is determined according to Figure 5.8. In addition the percentage of ships in each class, out of the total number of ships, is given in Table 5.2.

As calculated earlier, it is given that if an anchor crosses the pipeline, damage to the pipeline will occur only in 10% of the situations. This will determine the required soil cover on the pipeline.

From Table 5.2 it can be seen that 10% of all ships have an anchor penetration greater than 1.7 m. Consequently it can be concluded that the pipeline should have

Table 5.2   Assumed distribution of ship classes

Class	DWT	Anchor weight (tonnes)	Anchor penetration (m)	Percentage of ships
1	5 000	3	1.0	50
2	20 000	5	1.7	40
3	50 000	10	2.0	10

a soil cover of 1.7 m to fulfil the acceptance criterion. If the acceptance criterion is changed, this will change the requirements for soil cover on the pipeline. Similarly, a change in the ship traffic distribution will change the required soil cover.

5.6 Special risk aspects

5.6.1 *Subsea isolation valves*

In the event of an accident (e.g. fire) on a platform, the inventory of any incoming (import) or outgoing (export) pipelines may be prevented from reaching the platform by valves on the seabed. A possible solution for export pipelines is to install a valve assembly consisting of a check valve and a ball valve. In the case of loss of pressure on the platform side, the check valve would automatically shut down the backflow towards the platform, and tightness would be ensured by closure of the ball valve, either by a diver, by an ROV, or by a remotely operated actuator. The sequence of the valves is immaterial from a safety point of view, but for operational reasons (see Chapter 11) the check valve is placed closest to the platform.

Recently, it has become customary to replace the assembly by a single, remotely controlled subsea isolation valve (SSIV), which can also be installed to isolate the platform from an import pipeline. A subsea isolation valve with a protective structure is shown in Figure 5.9.

Since the accident on Piper Alpha in 1988, caused by a continuous and uncontrolled release of high pressure gas, it has been discussed whether or not to require the installation of SSIVs. The decision can be taken based on a full risk assessment of the problem. An important result of such a risk analysis will, in addition to the need for an SSIV, be the determination of the distance from the platform to the location of the valve, where two opposite effects must be considered. If the valve is located close to the platform, should there be damage the amount of medium released would be limited. On the other hand, the risk of damage of the pipeline on the 'wrong side' of the valve will increase. To determine the optimum distance from the platform to an SSIV the frequency of damage of the riser pipeline must be assessed. Hazards that could cause damage to the pipeline/risers are ship impact, fire on the platform, dropped objects, etc. That uncontrolled fire is still a very real risk is illustrated by the recent blaze on the Temseh platform in the Mediterranean, approximately 40 km off Egypt. Fortunately, all personnel were safely evacuated, but damage to equipment was considerable.

Figure 5.9   Subsea isolation valve with protection structure

5.6.2   *Welded or flanged connections*

An offshore pipeline is usually fully welded, except in deep water. However, close to platforms where risers, expansion offsets, valves or similar are installed, it is sometimes proposed that flanged connections are used as an alternative to welding. The decision between a flanged or a fully welded solution can be based on a risk assessment, including a cost benefit analysis. In this risk assessment both the installation and the operational phase should be considered.

In the operational situation it is assumed that the risk of product leak is higher for a flanged connection than for a fully welded system. However, the cost of a hyperbaric weld will be significantly higher than that of a flanged connection. Thus the decision on choice of a flanged or a welded connection can be taken following the principles for a cost benefit analysis as described in Section 5.4.2.

5.6.3   *Corrosion in HP/HT pipelines*

For pipelines operating at elevated pressures and temperatures internal corrosion may be a severe problem when water is present in the stream, which is typically the case for production flowlines. In this situation it is important to establish a well-documented corrosion management strategy (see Chapter 11). The corrosion management strategy should be based on the findings from the hazard identification.

5.7 Statistical data

To perform reliable risk assessments it is important to obtain, to the largest possible extent, reliable statistical data. Different types of statistics on failures, accidents and incidents are available from various sources, such as:

- PARLOC;
- E&P Forum;
- OREDA.

PARLOC is a database containing 'Loss of containment data for offshore pipelines', including information from regulatory authorities and pipeline owners. The database is updated at regular intervals, and the abbreviation stems from the first (1990) revision entitled *Pipeline and riser loss of containment study*.

E&P Forum is the Oil Industry Exploration and Production Forum, which is an international association of oil companies and petroleum industry organisations, formed in 1974. The organisation has prepared a quantitative risk assessment datasheet directory.

OREDA (Offshore Reliability Data) is a project organisation sponsored by eight oil companies with worldwide operations. Its main purpose is to collect and exchange reliability data among the participating companies, and to act as the forum for co-ordination and management of reliability data collection within the oil and gas industry (see www.oreda.com).

For onshore pipelines in Europe, spillage statistics are compiled by CONCAWE (www.concawe.be) for liquid pipelines, and by the European Gas Incident Group (EGIG) for gas pipelines.

However, it must be emphasised that in risk assessment there will always be some uncertainties connected with the results, and often sensitivity analysis should be performed to justify the conclusions.

Chapter 6
Design

6.1 Design conditions

6.1.1 *Codes and standards*

The client and authorities in the country where a pipeline is to be installed shall endorse the codes and standards used by the designer. Pipeline design codes that are widely recognised include:

- ASME B31.8-1999 Chapter VIII;
- BS 8010 Part 3;
- ISO 13623;
- DNV OS-F101.

A large number of pipelines have been and are successfully designed to the above codes. The list is by no means complete, and later in this chapter reference is made to Germanischer Lloyd (GL) and Deutsches Institut für Normung (DIN) codes.

Recently, a slightly amended ISO 13623 has been adopted by CEN as EN 14161 and, the British Standards Institution being a member of CEN, this means that BS 8010 will be withdrawn. However, the British national foreword to EN 14161 will advise that a more comprehensive approach to pipeline design is achieved by using the standard in association with PD (Published Document) 8010, which is an updated version of BS 8010.

In this book, most attention is given to the DNV code: first because this code governs in the North Sea area and secondly because the code has had international approbation. The DNV code is therefore considered the most appropriate standard for a future design.

ASME B31.8, BS 8010 Part 3 and ISO 13623 are all codes that belong to the Allowable Stress Design (ASD) family of codes. DNV OS-F101 adopts the Load and Resistance Factor Design (LRFD) format as a basis for the given structural limitations.

The traditional design of pipelines, where load factors of 0.5 and 0.72 typically have been used in the design of pipe wall thickness, exemplifies an allowable stress design (ASD) format. Designing a pipeline code using ASD is quite common, and parallels to the 'limit states' introduced below can be found.

For the determination of pipeline wall thickness the ASD and the LRFD formulation is identical. Sections 6.2.1 and 6.2.2 describe the wall thickness determination using hoop stress formulation and load factor design methods.

In Section 6.2.5 a comparison between DNV, ISO, GL and DIN codes is made. The principles for this comparison, regarding equalised load and resistance formulations, is an example only and could as well have been made for other recognised codes.

The use of limit states in the LRFD format is detailed in the subsections below.

Traditionally the following different limit states are considered:

- Serviceability Limit States (SLS);
- Ultimate Limit States (ULS);
- Accidental Limit States (ALS).

The design of the pipeline is closely related to the risk analysis, in the sense that scenarios that entail a risk that is unacceptable, typically due to their high frequency of occurrence, shall be considered in the ALS design. In this book, however, these cases, sometimes referred to as Accidental Load Situations, are included under ULS. Repeated loading is also considered as a ULS case, although some guidance documents (notably DNV OS-F101) identify a special Fatigue Limit State (FLS).

6.1.2 *Serviceability limit states (SLS)*

For a marine pipeline it shall be ensured that during its installation and operation it will not be unsuitable for its intended purpose. The SLS refers to a given load condition that, if exceeded, can cause the pipeline to be unsuitable for continued operation. The SLS are defined for all the relevant loading conditions that can be formulated. The following issues are normally considered:

- deformation and movements due to waves and currents (hydrodynamic stability);
- longitudinal deformations due to temperature and pressure variations (pipeline expansion);
- lateral deformations due to restrained temperature and pressure expansion (upheaval buckling or snaking);
- blockage of the pipeline, due to hydrate formation or wax deposition (flow assurance), for example.

A distinction is made between reversible and irreversible SLS. In the case of permanent local damage or permanent unacceptable deformations the SLS is not the appropriate formulation and the ULS design factors shall be introduced. This implies that when first time occurrence constitutes failure the formulation would be ULS.

For many SLS the transition from the desired state to the undesired state is rather vague; the transition implying a more or less slowly decreasing degree of serviceability.

6.1.3 *Ultimate limit states (ULS)*

It shall be ensured that the pipeline has the required safety against failure in the ULS, defined in terms of:

- plastic deformations (yielding);
- local instability (buckling);
- crack instability (bursting);
- repeated loading (fatigue).

Furthermore, it shall be ensured that the pipeline has the required safety against accidental loads.

The design criteria for ULS should in general be formulated for the first passage of the limit state, as the first passage in almost all cases is equivalent to failure. Note, however, that yielding failure is defined in terms of deformations, not stresses, implying that first time yield is allowed, provided it does not lead to excessive strains or deformations.

It should also be noted that fatigue or other time dependent deterioration mechanisms reduce the strength of the structure, and may initiate ULS. In this relation it is useful to distinguish between damage tolerant and damage intolerant structures. For the latter, fatigue may be treated as an ULS, whereas it may be considered as an SLS for a damage tolerant structure.

6.1.4 *Partial safety coefficients*

To ensure the required safety against the defined limit states it is customary to use the partial safety coefficient method, which requires that a number of partial safety coefficients be defined for both actions and resistance parameters.

The partial safety coefficient reflects the uncertainty of the parameter in question, and it is assigned a numerical value greater than unity, which is used as a multiplier on the actions, and as a divisor on the resistances. Thus a large factor corresponds to a low level of certainty. Ideally, the partial safety system should be calibrated to give a uniform safety against all possible failure scenarios, expressed by a target safety level (reliability index β). In reality the adopted coefficients reflect accepted practice and national priorities. In pipeline design the inverse of the partial safety coefficient on the linepipe steel strength is traditionally termed the usage factor or utilisation ratio, which thus is a factor of less than unity.

The required safety level (pipeline safety class) will normally depend upon the consequences of pipeline rupture, and it is customary to divide the route into zones with different safety classes, see Section 6.2.3.

If required it will be possible to make a code calibration of the partial safety coefficients. Such a calibration shall be based on a specified value of the acceptable frequency for a severe damage of the pipeline.

6.2 Wall thickness determination

6.2.1 *General*

The primary objective of the linepipe design is to determine the optimal wall thickness and steel grade of the pipeline. For the vast majority of existing pipelines the wall thickness will have been selected following a simple hoop stress calculation. A usage factor applied to the Specified Minimum Yield Stress (SMYS) defines the allowable stress which, when inserted into the hoop stress formulae, determines the minimum required thickness of the pipe wall. The engineer would then select the nearest standard API wall thickness above the required minimum. For liquid or two-phase pipelines a corrosion allowance may have been added. There are numerous approaches to determining the corrosion allowance; see Section 6.6.3. Calculations that aim at a specified design life are often backed up by extensive testing, but the corrosion allowance may also simply be based on experience with existing lines or on owner preferences.

The engineering of an optimum wall thickness and steel grade can be much more complicated, involving requirements to steel mills for steel quality that follow strict statistical distributions, and involving integrated design where the pressure safety system trip accuracy at inlet facilities is ensured to function within a given range. Additional complications are introduced if the resulting wall thickness appears uncomfortably thin, normally when the D/t ratio exceeds approximately 45.

In technical terms the linepipe design covers the determination of the wall thickness and steel grade required to contain the internal pressure, as well as identification of where and to what dimensions excess wall thickness must be selected. Apart from the corrosion allowance discussed above, excess wall thickness may follow from analyses of loads during installation (e.g. reeling), or it may be a requirement arising at a particular point of concern (e.g. cable crossing). The interest in optimising the pipeline wall thickness is particularly obvious for large transmission lines, typically gas pipelines. The cost of the bare steel pipe may be up to 50% of the entire pipeline project cost, and price tags counted in hundreds of millions USD have been seen for a number of recent projects. The steel grade selection includes not only the determination of specified minimum yield stress, but also all metallurgical considerations that eventually go into the preparation of the linepipe purchase specification.

When designing a pipeline the pressure value in focus has traditionally been the design pressure. Though this value is obviously relevant for design, in recent years there has been a growing understanding of the importance of the definition, detailing and control of the pressure loads that the pipeline will actually experience during service. In the context of wall thickness design the most direct cost optimisation (often tediously labelled 'savings'), gained from code developments in recent years, is related to the correct design of the pressure control system. Pressure control systems are discussed in Section 4.2.3, and the example included gives a detailed insight into the practical application, as well as to the interface to land pipeline systems at the onshore terminal. With modern code practice, specifically the DNV OS-F101 standard, the design pressure is essentially a nominal value used for design, determined by the physical and actual pressures, namely maximum operating and incidental pressures. Alternatively, the design pressure may be a starting point, determining the maximum operating and incidental pressures, which then govern the design of the pressure control system.

Other loading cases may influence the design of the pipe wall thickness. These include installation loading, bending loads due to seabed undulations, crossings, etc., as well as external impact (e.g. from trawling) or loading from the external over-pressure caused by the water column above the line.

The wall thickness design may result in a pipe that is too thin-walled for practical use, or rather for comfort. DNV has a requirement of minimum 12.0 mm wall thickness for pipes larger than 8″ diameter in high safety class and Location Class 2 (see Section 6.2.3).

6.2.2   *Design methods*

Wall thickness design according to various codes always focuses on (variations of) the hoop stress formula, where a stress ratio against SMYS is easily understood, and hence the code expression would appear clear. In reality, comparison of different codes is very complicated. The combination of varied definitions of pressures combined with varying load factors (usage factors), steel specifications and code particulars, stress definitions, safety class denominations, etc. leads one to the conclusion that the calculation of a suitable wall thickness is far from being a trivial task. A wall thickness design code comparison sample exercise is presented in Section 6.2.5.

The traditional hoop stress formula would read:

$$t \geq \frac{\Delta p \cdot D_o}{2 \cdot SMYS} \cdot \gamma_1 \cdot \gamma_2$$

where

t	pipe minimum wall thickness
Δp	internal overpressure
D_o	pipe diameter, average or outer
SMYS	specified minimum yield stress of motherpipe material
γ_1	Safety factor No 1, e.g. stress utilisation
γ_2	Safety factor No 2, e.g. related to materials.

The hoop stress is the dominant, but not the only, stress component: the pipe wall of a pipeline in operation may also be subjected to axial stresses induced by pressure and temperature, as well as occasional bending stresses; see Sections 6.4.4 and 6.5. A state of plane stress is assumed, and the verification is based upon a suitable yield condition, normally Tresca or von Mises. For a given uniaxial yield stress (SMYS) the former is slightly more conservative than the latter, and this may be reflected in the choice of outer or average diameter in the above hoop stress formula.

6.2.3 *Location class definition*

Design of the pipeline wall thickness is based on the classification of the pipeline into safety classes, based on location and transported medium. The potential consequences of failure are addressed with the human density as the dominant factor. Codes distinguish between geographical zones, and the DNV zone terminology uses two location classes, defined below.

Location Class 1
Locations where no frequent human activity is anticipated. For most offshore pipelines there is little risk of human injury in case of failure, and the majority of the pipeline route is therefore normally assigned to Location Class 1.

Location Class 2
Locations near (manned) platforms or areas with frequent human activity (e.g. landfalls). The extent of Location Class 2 from these areas is typically a minimum distance of 500 m.

The extent of the Location Class 2 would normally be reckoned from the centre, or the closest accommodation module, of the platform, but DNV has clarified that the 500 m should be taken from the riser touch-down point on the seabed.

Another definition given in ISO 13623 uses five location classes, covering both onshore and offshore pipelines. Basically, marine pipelines correspond to ISO 13623 Location Class 1 and 2.

The adoption of DNV Location Class 1 offshore is the normal practice. Offshore human density assessments, taking into account the number of ships passing a

Table 6.1 Typical safety classes

Phase	Location Class	
	1	2
Installation	Low	Low
Operation	Normal	High

given area, can be considered analyses of human activity. The outcome 'frequent human activity' might result, but in reality even heavy ship traffic normally does not translate into a permanent population density warranting use of Location Class 2. There are of course special considerations, such as harbour areas, offshore mining or intense ferry traffic, that could alter this conclusion.

The designation of safety classes also involves a classification of the transported medium. The fluid category with regard to toxicity and environmental impact upon release needs addressing. For instance, one fluid category covers non-toxic, single phase gas, which is mainly methane, and is hence non-toxic, but flammable.

Considerations of contingencies and security of supply in case of major damage or rupture will sometimes be taken into account when the pipeline safety class is determined. Pipeline connections creating redundancy in the supply, or storage facilities, will normally be considered to constitute acceptable redundancy, in which case the pipeline will not be designed according to 'High Safety Class', but according to normal pipeline design practice.

Concerning the basis for derivation of safety class, location class and fluid classification, reference is made to DNV OS-F101 Section 2. Using this code the resulting safety class (Low, Normal or High) impacts the safety factors, and therefore the required wall thickness. As shown in Table 6.1, the resulting classification is also dependent on the pipeline use.

6.2.4 *Wall thickness according to DNV OS-F101*

DNV OS-F101 is based on an LRFD (Load and Resistance Factor Design) format, i.e. load factors are applied to the loads, and a resistance factor is applied to the material strength. The nomenclature from DNV OS-F101 is used in the following.

The following equation shall be satisfied:

$$p_{li} - p_e \leq \frac{p_b(t_1)}{\gamma_{SC} \cdot \gamma_m}$$

where

p_{li} local incidental pressure

p_e external pressure

$p_b(t_1)$ pressure containment resistance

t_1 minimum wall thickness

γ_{SC} safety class resistance factor

γ_m material resistance factor.

The local incidental pressure is given by

$$p_{li} = p_d \cdot \gamma_{inc} + \rho_{cont} \cdot g \cdot H$$

where

p_d design pressure at reference point

γ_{inc} incidental pressure to design pressure ratio

ρ_{cont} density of content

g acceleration of gravity

H height difference between chosen point and reference point.

The factor γ_{inc} shall take a minimum value of 1.05. In DNV OS-F101 Section 3 B305, this is allowable provided the pressure safety system is specified to ensure that the local incidental pressure cannot be exceeded. This is ensured following the guidelines on pressure control systems as given in Section 4.2.3. The maximum value of $p_{li} - p_e$ governs the wall thickness. The maximum wall thickness is found at mean sea level, where the external pressure equals zero. As the water density is greater than the density of the product the value of $p_{li} - p_e$ decreases with increasing water depth, but normally the difference is not sufficient to justify a jump in wall thickness.

The basis for derivation of safety class, location factor and fluid classification according to DNV OS-F101 Section 2 is presented in the section above.

The safety class resistance factor is given in Table 6.2 (see DNV OS-F101 Section 5 D207).

The wall thickness calculations are then based on normal safety class for offshore section (Location Class 1) and high safety class (Location Class 2) for landfalls or platform proximity areas.

The material resistance factor γ_m shall for limit state categories SLS (Serviceability Limit State), ULS (Ultimate Limit State) and ALS (Accidental Limit

Table 6.2   Safety class resistance factor γ_{SC} for pressure containment

Safety class	Low	Normal	High
γ_{SC} (hoop)	1.046	1.138	1.308

State or Accidental Load Situation) be taken as 1.15, see DNV OS F-101 Section 5 D206.

The pressure containment resistance, $p_b(t_1)$, is given by:

$$p_b(t_1) = \min\{p_{b,s}(t_1), p_{b,u}(t_1)\}$$

where

$p_{b,s}(t_1)$ pressure containment resistance for yielding limit state
$p_{b,u}(t_1)$ pressure containment resistance for bursting limit state.

The yielding limit state is:

$$p_{b,s}(t_1) = \frac{2 \cdot t_1}{D - t_1} \cdot f_y \cdot \frac{2}{\sqrt{3}}$$

The bursting limit state is:

$$p_{b,u}(t_1) = \frac{2 \cdot t_1}{D - t_1} \cdot \frac{f_u}{1.15} \cdot \frac{2}{\sqrt{3}}$$

where

f_y design yield stress
f_u design tensile strength
D nominal outer diameter.

The design yield stress and the design tensile strength can be found as:

$$f_y = (SMYS - f_{y,temp}) \cdot \alpha_U$$
$$f_u = (SMTS - f_{u,temp}) \cdot \alpha_U \cdot \alpha_A$$

where

$f_{y,temp}$ temperature de-rating value, 0 if the design temperature is below 50°C

$f_{u,temp}$ temperature de-rating value, 0 if the design temperature is below 50°C

α_U material strength factor, 0.96 for normal and 1.00 for supplementary requirements, suffix U

α_A anisotropy factor, 1.00 for the circumferential direction

$SMYS$ Specified Minimum Yield Stress

$SMTS$ Specified Minimum Tensile Strength.

For pipelines operated at high temperatures (i.e. above 50°C for carbon steel) the yield strength will thus be reduced according to DNV OS-F101.

6.2.5   *Code comparison and national wall thickness regulations*

The fundamental definition of pressure and control has a major impact on the overall safety level of the design. The illustrative example described below is based upon a specific project, set in a specific geographic context, but the example sheds light upon the wall thickness determination using various design standards. The comparison covers the following codes:

DNV   Det Norske Veritas, Offshore Standard OS–F101, *Submarine pipeline systems*, 2000.

ISO   International Standard ISO 13623, *Petroleum and natural gas industries – Pipeline transportation systems*, 2000.

GL   Germanischer Lloyd, *Rules for classification and construction, III – Offshore technology, Part 4 – Subsea pipelines and risers*, 1995.

DIN   German Standards Committee, DIN 2413, *Steel pipes, calculation of wall thickness subjected to internal pressure* (1972) October 1993.

The inclusion of the German DIN 2413, when there exists a German code (GL) specifically for offshore use, is because of practical examples where offshore pipelines, for the wall thickness design, are required to be analysed against the landline code (DIN). This again serves to highlight that the wall thickness determination, particularly for major gas trunk lines, can be a major issue for debate between the designer, authorities, certifying agency, etc. Well proven as land codes are in most regions of the world, there are plentiful bases for comparison if the designer tries to introduce a 'new' set of rules. The following therefore serves partly to highlight the differences, or the fact that these are limited, as well as to give an example of how a code calibration can be undertaken.

The starting point in a calibration exercise is to try to examine what the safety philosophy behind individual code requirements is.

Pressure definitions

DNV Section 12 is commentary (information) material, where amongst other things the basic principles of the design are described. This section would therefore serve as an indication for the adopted safety philosophy.

DNV OS-F101 Section 12 E 200 *Conversion of pressures* explains the design philosophy as follows:

> The governing pressure for design is the incidental pressure. The incidental pressure is normally defined as the pressure with an annual probability of exceedance of 10^{-4}.

If the design pressure is given, the incidental pressure shall be determined based on the pressure regulating system and pressure safety system tolerances and capabilities to ensure that the local incidental pressure meets the given annual probability of exceedance above.

The DNV code further states that the incidental pressure minimum shall be taken as 5% above the design pressure, and a maximum of 10% above the design pressure. A situation where the incidental pressure is more than 10% above the design pressure would lead to redefinition of the design pressure.

It is of paramount importance to understand these terms defining pressure values. Figure 4.1 reproduces the pressure definitions from DNV. The calibration of codes can only meaningfully be conducted if a normalisation of the terms is attempted. As will be shown, there appears to be significant divergence between the DNV, ISO, GL and DIN codes (note that the latter does not specifically refer to offshore pipelines).

The difference between the Maximum Allowable Operating Pressure (MAOP) set point and the design pressure is normally taken as 2%.

Code comparison – utilisation ratios

The DNV utilisation ratios together with the comparative numbers from the ISO, GL and DIN codes are given in the Table 6.3. The usage factors are defined as the hoop stress to yield stress ratio, where the hoop stress is based on the mean diameter. This is important, as the Germanischer Lloyd and DIN 2413 codes employ outer diameter in the wall thickness determination. Furthermore, the

Table 6.3 Hoop stress design factors – nominal comparison for fixed design pressure

Code comparison design factors	Usage factor for pressure containment (made equivalent to hoop stress formulation based on mean diameter)	
Design Code	Location Class 1 – normal safety class (away from frequent human activity)	Location Class 2 – high safety class (landfalls, crossings, structures)
DNV OS-F101, with steel requirements suffix U	0.802*/0.840[†]	0.698*/0.731[†]
DNV OS-F101, without steel requirements suffix U	0.77*/0.807[†]	0.67*/0.702[†]
ISO 13623	0.77	0.67
Germanischer Lloyd	0.783	0.555
DIN 2413	0.703	0.703

* Incidental pressure to design pressure ratio equal to 1.10
[†] Incidental pressure to design pressure ratio equal to 1.05

basis for the calculations is a fixed design pressure, which is used explicitly in wall thickness calculations in ISO 13623, Germanischer Lloyd and DIN 2413. For the two checks to DNV OS-F101 the design pressure is implicitly taken as incidental pressure divided by 1.10 and 1.05, respectively.

Conclusion on utilisation ratios

For identical pressure definitions, the DNV, ISO and GL design codes are concluded to produce almost identical hoop stress usage levels, albeit the GL code prescribes slightly higher usage in Location Class 1, and lower in Location Class 2. The DIN code, on the basis of identical design pressure values, prescribes lower utilisation ratios in Location Class 1.

Code comparison – design pressure definitions

The above discrepancies in utilisation ratios appear partly because comparison is made on the basis of a fixed design pressure. Since the safety level is closely related to the incidental pressure (or peak pressure as used in some codes) a consistent safety level is not achieved in such a comparison. Instead a fixed incidental pressure should be employed as the basis for the comparative calculations.

A HIPPS assisted pressure relief system warrants the use of a maximum incidental pressure of 1.05 times design pressure; see Section 4.2.3. Hence, introducing a numerical example using a design pressure of 180 barg, the incidental pressure is 1.05×180 barg $= 189$ barg. Based on this value, equivalent design pressures are calculated from the given ratios between incidental pressure and design pressure found in the various codes, see Table 6.4.

It is apparent from Table 6.3 that the inherent safety level in ISO 13623, Germanischer Lloyd and DIN 2413 is greater than that in DNV OS-F101 when the design pressure is the point of reference and the minimum ratio of 1.05 between incidental pressure and design pressure is used in DNV OS-F101.

However, making recalculations based on the listed equivalent design pressures, a more consistent safety level among the codes is obtained. The resulting nominal wall thicknesses are listed Table 6.4. It is seen that close agreement is found between DNV OS-F101 (without requirement U, which is a 'high utilisation' designation referring to a set of requirements to the steel quality, specifically mill test results for SMYS in the transverse direction), ISO 13623 and Germanischer Lloyd in Location Class 1, whereas in Location Class 2 Germanischer Lloyd is significantly more conservative. The code DIN 2413 is the most conservative in Location Class 1, but is in close agreement with DNV OS-F101 and ISO 13623 in Location Class 2.

Conclusion on design pressure definitions

When identical pressure definitions are adopted, the international codes DNV, ISO and GL all yield similar pipe wall thicknesses.

Table 6.4 Wall thickness design for 28″ line – calibrated comparison for fixed incidental pressure

Code comparison design wall thickness	Incidental pressure to design pressure ratio	Equivalent design pressure (barg)	Nominal wall thickness for pressure containment given 189 barg incidental pressure (made equivalent through code calibration) (mm)	
Design Code			Location Class 1	Location Class 2
DNV OS-F101, with steel requirements suffix U	1.05	180.0	17.7	20.3
DNV OS-F101, without steel requirements suffix U	1.05	180.0	18.5	21.1
ISO 13623	1.10	171.8	18.5	21.1
Germanischer Lloyd (including 1 mm corrosion allowance)	1.15	164.3	18.4	25.2
DIN 2413 (excluding 1 mm corrosion allowance)	1.00/1.20*	180.0/157.5	24.2/20.2	24.2/20.2

* DIN 2413 interpretation can be debated against very specific conditions.

Code comparison – pressure regulation

Further to the discussion of safety level, it is worth considering the inclusion in DNV of a set point tolerance, which is typically 2%. The ISO 13623:2000E states in Clause 6.3.2.2 that:

> the internal design pressure at any point in the pipeline shall be equal to or greater than the maximum allowable operating pressure (MAOP).

An identical formulation can be found in Germanischer Lloyd Clause 4.2.1.2.

It could be argued that the set point tolerance should be deducted irrespective of the above statement. This is not warranted, however. In codes that allow the MAOP to be equal to the design pressure the set point tolerance must be assumed inherent in the wall thickness design factors, and hence the safety philosophy must be assumed to reflect this. The singling out of the set point tolerance is a relatively new feature from DNV, and there are numerous pipelines in service following the earlier design standards, for example DNV 1981, ASME B31.3-2002 and BS 8010, where the design pressure and the maximum operating

Table 6.5 Code comparison – governing pressures

Design code	Design pressure (barg)	Maximum operating pressure (barg)	Incidental pressure code (barg)
DNV OS-F101	180.0	176.5	189.0
ISO 13623	180.0	180.0	198.0
Germanischer Lloyd	180.0	180.0	207.0
DIN 2413	180.0	180.0	216.0

pressures are equal. These codes specify a usage factor of 0.72, and the development therefore appears as consistent.

Table 6.5 summarises the various pressures governing the wall thickness selection following the codes in question.

It is seen that the apparently higher utilisation allowed by the DNV OS-F101 is based on tighter requirements to pressure regulation system.

Conclusion

In the preceding sections, it has been demonstrated that international codes under identical definition of loading (pressure) would prescribe almost identical pipe wall thicknesses.

6.2.6 *Trawling and hydrostatic pressure*

Some design considerations are presented below for deformation of the cross-section due to:

- trawl impact;
- trawl hooking and pull-over;
- external pressure only (collapse);
- propagation buckling.

In all the subsequent formulae the following apply:

D nominal outer steel pipe diameter
t nominal steel wall thickness
f_y design yield stress (see Section 6.2.4)
γ_m material resistance factor, taken at $\gamma_m = 1.15$
α_{fab} fabrication factor, depending upon the pipe manufacture:
           seamless: $\alpha_{fab} = 1.00$
           welded and expanded: $\alpha_{fab} = 0.85$

Trawl impact

The design criterion for trawl impact is related to the acceptable dent size, and design guidance may be found in the DNV Guideline No 13 *Interference between trawl gear and pipelines*. The acceptable dent depth is expressed as:

Table 6.6 Acceptable dent size depending on trawling frequency

Frequency class	Impact frequency (per km per year)	Usage factor, η	Dent size, H_p/D (%)
High	>100	0.0	0.0
Medium	1–100	0.3	1.5
Low	<1	0.7	3.5

$$H_p/D = 0.05\eta$$

where

H_p characteristic permanent plastic dent depth
η usage factor.

The usage factor, determining the acceptable dent size, depends upon the trawling frequency, and may be taken from Table 6.6.

Guidelines for the calculation of dent depths are given in the DNV Guideline No 13, which also specifies reduction factors, depending upon soil conditions and pipe diameter, that may be applied to the trawl board steel mass and added mass. Ultimately, more detailed FEM analyses may be needed. The dent values given in Table 6.6 are only directly valid for the trawl load case, thus a larger dent could be found acceptable for a different accidental load, e.g. anchor impact.

Normal anti-corrosion or insulation coatings do not offer any significant protection against trawl impact, but a concrete coating may be assumed to absorb an impact energy of the order of 5.4 kJ.

Trawl hooking and pull-over
Hooking is a situation where the trawl gear gets stuck under the pipeline, subjecting the line to a force as large as the breaking strength of the trawl warp line (see Figure 4.4). It is normally considered an accidental load situation, implying that all load and resistance factors are set at unity. Hooking is a risk at free spans of the pipeline, and the DNV Guideline No 13 considers two cases:

(1) part penetration (may occur for all span heights);
(2) wedging (may occur for free spans above a critical height H_{cr}).

The critical height depends upon the type of trawl gear, and DNV Guideline No 13 gives the following value for trawl boards:

$$H_{cr} = 0.35H_B$$

where

H_B height of the trawl board.

The hooking acceptance criteria are related to the pipe resistance to local buckling, as the pipeline is lifted off the seabed to a maximum height H_1, whereupon the trawl board cuts loose, and the response can be calculated by a static analysis, applying the maximum lifting height as a prescribed deflection. The lifting height H_1 depends on the hooking case, and DNV Guideline No 13 gives the values:

$$H_1 = 0.35H_{\mathrm{B}} - 0.3D_{\mathrm{o}} \quad \text{(part penetration)}$$

$$H_1 = 0.5H_{\mathrm{B}} \qquad\qquad \text{(wedging)}$$

In most cases, however, it will be the pull-over forces (see Section 4.4.3) that will limit the allowable span height. Thus the question to investigate is whether there is topical trawl gear that is small enough to get hooked by the spans allowed by the pull-over analysis, and yet sufficiently strong to damage the pipeline.

System collapse

System collapse refers to flattening of the pipe profile due to external hydrostatic overpressure. The flattening can be initiated if the pipeline locally is deformed to an oval shape. The deformation could stem from an incident during laying giving a large point load at the outer roller support, or bending moments in a pipe spanning a seabed obstacle, rock outcrop or similar.

The characteristic resistance for external pressure p_c or collapse according to DNV OS-F101, is calculated from:

$$(p_c - p_{el}) \cdot (p_c^2 - p_p^2) = p_c p_{el} p_p \Delta D \left(\frac{D}{t_2} \right)$$

where

p_c characteristic collapse pressure

p_{el} elastic collapse pressure:

$$p_{el} = \frac{2E}{1 - v^2} \left(\frac{t_2}{D} \right)^3$$

p_p plastic collapse pressure, defined as

$$p_p = 2 f_y \alpha_{\mathrm{fab}} \left(\frac{t_2}{D} \right)$$

t_2 pressure resisting thickness, i.e. thickness reduced by the corrosion allowance:

$$t_2 = t - t_{corr}$$

except during installation and pressure testing

ΔD ovality, defined as

$$(D_{max} - D_{min})/D$$

In the absence of documentation the ovality is taken as 2%, i.e. $\Delta D = 0.02$, and should at minimum be assumed at 0.005.

The DNV collapse system check requires that

$$p_e \leq \frac{p_c}{1.1 \gamma_m \gamma_{SC}}$$

The safety class resistance factor γ_{SC} is given in Table 6.7.

Propagation buckling

Propagation buckling is the flattening of a large section of pipeline due to external pressure alone. Once local buckling has occurred, for whatever reason, if the external pressure exceeds a certain value, the buckle initiation pressure p_{init}, a propagating buckle will form, and will travel along the pipeline until the pressure is reduced to a lower value, the minimum propagation buckling pressure p_{pr}.

According to DNV OS-F101, the expression for the propagation buckling pressure is

$$P_{pr}(t_2) = 35 \frac{f_y \alpha_{fab}}{\gamma_m \gamma_{SC}} \left(\frac{t_2}{D}\right)^{2.5}$$

where γ_{SC} shall be taken from Table 6.7.

Table 6.7   Safety class resistance factor γ_{SC} for buckling

Safety class	Low	Normal	High
γ_{SC}	1.04	1.14	1.26

Summary of buckling criteria and design

- The collapse pressure p_c is the external pressure required to buckle a pipeline due to external pressure (and ovality) alone.
- The initiation pressure p_{init} is the external pressure required to start a propagating buckle from a given buckle. This pressure will depend on the size of the initial buckle.

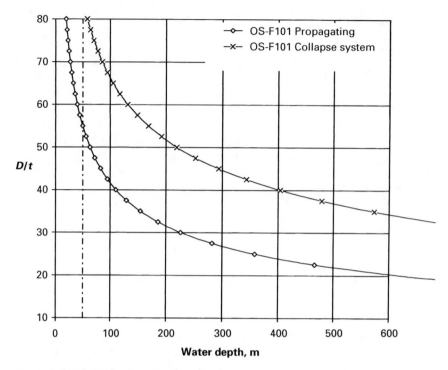

Figure 6.1 Buckling check for external pressure – vertical dashed line is an example water depth

- The propagation pressure p_{pr} is the external pressure required to continue a propagating buckle. A propagating buckle will stop when the external pressure is less than the propagation pressure.

Propagating buckling cannot be initiated unless local buckling has occurred. The relation between the different pressures is

$$P_c > p_{init} > p_{pr}$$

Figure 6.1 shows the collapse system check and propagating pressure expressed in terms of water depth as function of diameter-to-thickness (D/t) ratio.

The design should ensure a pipe wall thickness where the system collapse criterion is fulfilled at all times. However, it is normal design practice to perform the system collapse check for the installation phase only, which implies that any corrosion allowance is included in the wall thickness. The reasoning is that once the pipeline is in operation it is pressurised, and thus not susceptible to collapse buckling.

When, during the design, the wall thickness computes at a D/t ratio within the propagating buckle area careful considerations into safety level and buckling counter measures should be initiated.

The normal design remedy is to introduce buckle arrestors, i.e. short sections of heavy wall pipe with a resistance that is sufficient to prevent a buckle, once initiated, from flattening large sections of the pipeline. Buckle arrestors may take the form of integrated ring stiffeners, welded into the line, or of single, oversized pipe joints. The actual design and the spacing of these arrestors can follow safety studies, including evaluation of buckle initiation failures.

Various design procedures can be used, see, for example, DNV OS-F101 or Sriskandarajah and Mahendran (1987). Although the propagation buckling formula given above is quite conservative, DNV OS-F101 recommends introducing an additional safety factor of 1.5 (effectively reducing the factor 35 to 30) for the buckle arrestor pipe, due to the dynamic overpressure when a propagating buckle is abruptly stopped. The result of the analysis will be a range of feasible buckle arrestor lengths and thicknesses, the required thickness being constant once the length exceeds a few metres. However, due to the increased safety factor, this thickness will be larger than that required to resist propagation buckling of a continuous pipeline.

Often the length of Location Class 2 pipeline will be insufficient to justify the introduction of buckle arrestors and, in order to avoid more than two linepipe sizes, single Location Class 2 joints may conveniently be used as buckle arrestors on the remaining pipeline in Location Class 1. Thus in this case the buckle arrestor design in Location Class 1 will determine the wall thickness for Location Class 2.

The optimal spacing of the buckle arrestors may ideally be determined by a cost benefit analysis, weighing the total cost of the buckle arrestors against the risk of flattening a corresponding length of pipeline. The minimum is quite flat, however, and it is common practice to install single joint buckle arrestors spaced 500–600 m apart, corresponding to 40–50 pipe joints.

6.2.7 *Wall thickness design example*

Table 6.8 presents a design example where the required nominal wall thickness for a 28″ Location Class 1 offshore pipeline is given for the various loading conditions, reflecting installation procedures described in Chapter 8. It is seen

Table 6.8 Required nominal wall thickness based on Grade X65 steel for 28″ Location Class 1 offshore pipeline

Load condition	Req., *t* (mm)
Int. pressure according to DNV	15.3
Buckling due to external pressure	<15.3
Stacking analysis	<15.3
Laying analysis	<15.3
Trenching analysis	16.4

Table 6.9 Selected pipe dimensions for the offshore pipeline

Location	Approximate length	D (mm)	t (mm)	D/t ratio
Location Class 1	300 km	711	16.4	43.3
Location Class 2	3 km	719	20.5	35.1
Near shore section	5 km	719	20.5	35.1
Tie-in spools	0.6 km	726	24.1	30.1

from Table 6.8 that with respect to buckling and laying conditions the required wall thicknesses are all less than that required for the pressure containment. However, a trenching analysis shows that increased wall thickness is required to avoid overstressing the pipeline. Thus unless another trenching method is chosen, the requirements for trenching will govern the wall thickness selection. The trenching analysis conducted is based on realistic assumptions of the trench sled support configuration and trench depth. The required nominal wall thickness has been calculated at 16.4 mm, which results in a system bore of 678.2 mm.

For the Location Class 2 areas, it has been decided for this particular project to introduce a more rigid requirement than could be deduced from the DNV standard. A maximum pipeline hoop stress utilisation ratio of 0.6 based on minimum wall thickness is adopted. The required nominal wall thickness has been calculated at 20.5 mm.

Analysis of a davit lift tie-in operation indicates that special precautions need to be taken to avoid overstressing the pipe at the lower bend. Adopting the Location Class 2 wall thickness of 20.5 mm over this short section reduces the stresses to a manageable level.

At platform and structure tie-in spools, at possible future hot-tap locations, and where pipes are to be welded to subsea valves, excess wall thickness is preferred. A wall thickness of 24.1 mm is selected for these applications.

Based on pressure containment requirements, buckling and installation analyses the selected wall thickness and corresponding outer diameter for the sample offshore pipeline are shown in Table 6.9.

6.3 Hydrodynamic stability

6.3.1 *General*

On-bottom stability analysis is performed to ensure the stability of the pipeline, when exposed to wave and current forces and other internal or external loads (e.g. buckling loads in curved pipeline sections). The requirement to the pipeline is that no lateral movements at all are accepted, or alternatively that certain limited movements that do not cause interference with adjacent objects or overstressing of the pipe are allowed.

Hydrodynamic stability is generally obtained by increasing the submerged weight of the pipe by concrete coating. Other means may sometimes be applied, such as increasing the steel wall thickness, placing concrete blankets or bitumen mattresses across the pipeline, anchoring, or covering it with rock or gravel. The design criterion applied shall of course reflect the actual method used for stabilisation.

Alternatively, the hydrodynamic forces may be reduced by placing the pipeline in a trench on the seabed, prior or subsequent to installation. Hydrodynamic stability ceases to be an issue if the pipeline is artificially or naturally backfilled. The natural backfilling of a trenched pipeline depends on the environmental conditions and the seabed sediment at the location. Natural backfilling assessment requires the calculation of sediment transport for a number of representative seastates, combined with the evaluation of the entrapment of the sediment in the trench. Natural variability of the seabed level has to be assessed for the determination of the required trenching depth and the time required for natural backfilling to take place. In a sandy seabed subjected to significant wave action and high rates of sediment transport it is also possible to rely on self-burial, i.e. the propensity of any object on the seabed to sink into the sand as a result of erosion and scour processes and natural backfilling.

A pipeline on the seabed forms a structural unit where displacements in one area are resisted by incurred bending and tensile stresses. Residual stresses from the laying process may also provide resistance against displacement. Although the term 'on the seabed' is applied, the real situation most probably involves a great variety of pipeline–seabed interface conditions. Pipeline self-lowering may result in some sections of a pipeline being embedded to a substantially larger degree than determined by touchdown forces, and parts may even be fully buried. The embedment is influenced by soil characteristics, as well as by phenomena such as scour, sediment transport, and other seabed instabilities. In other sections the pipe may be slightly elevated above the seabed due to seabed undulations or scour processes. For both conditions (embedded/buried or free spanning pipe) the hydrodynamic forces are reduced relative to the idealised on-bottom condition.

Soil resistance forces will also be heavily affected by embedded/buried (or spanning) pipe sections. In general, the actual soil resistance is a function of the load history, and it is larger for cyclic loading than for static, unidirectional loading. The soil resistance is often assumed to be made up of frictional forces determined by the effective weight of the pipeline (submerged weight minus lift force) and a passive soil resistance due to embedment. The soil resistance varies along the pipeline, and in the case of lateral pipe displacements, longitudinal soil resistance will also develop. The pipeline–seabed interaction is thus in general fairly complex, and requires quite detailed information and advanced methods and calculation tools if detailed analysis is applied.

Previously, on-bottom stability analysis has been performed using simple methods, representing the actual conditions in a relatively crude manner. Today more complex methods are applied as reflected in DNV's Recommended Practice, RP E305 *On-bottom stability design of submarine pipelines* or the PRCI Guideline, *Submarine pipeline on-bottom stability, analysis and design guidelines.*

6.3.2 *Design activities*

The calculation procedures for on-bottom stability include, in principle, the following steps:

- determination of near seabed flow conditions;
- determination of hydrodynamic forces and soil reaction forces;
- hydrodynamic stability check.

The near seabed flow conditions are characterised by:

- wave induced (orbital) water velocities and acceleration at the seabed;
- steady current velocity at the seabed.

It is common for on-bottom stability design procedures that only the flow and acceleration components perpendicular to the pipe axis are considered in the design process. The main directions for waves and current induced flow and possible directional spreading of the wave induced flow are therefore important to consider.

Orbital velocity and acceleration
The determination of the near seabed orbital velocity and acceleration can be based on a single wave transfer or on spectral transfer of a seastate. Both cases, however, require application of a wave theory.

Comparison between several wave theories and field and laboratory data has demonstrated that linear wave theory provides good predictions of near bottom kinematics for a fairly wide range of relative water depths and wave steepnesses. One reason for this relatively good agreement is that the influence of non-linearity is attenuated with depth below the free surface. Sinusoidal theory is not capable of describing the near seabed kinematics under breaking waves. In this situation a transportation of mass is initiated, and the phase relation between the velocity and acceleration is no longer $\pi/2$ due to the modified wave profile. In the case of shallow water close to the breaking zone, theories other than linear theory should be applied, for example, higher order Stokes or Stream Function theory. The importance of an accurate assessment of the wave induced bottom kinematics under breaking waves may be less significant in cases such as a shore approach perpendicular to a straight coastline, where the wave induced bottom velocity will be almost parallel to the pipeline as a result of wave refraction.

The graph in Figure 6.2 could support the decision to use non-linear wave theory.

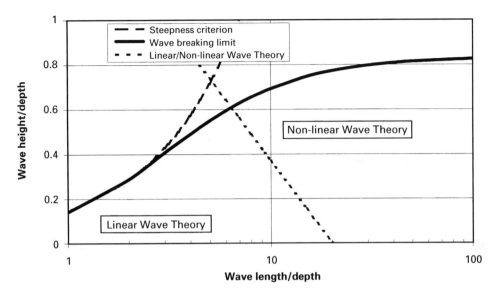

Figure 6.2 Wave theory application

First order wave theory is described in many textbooks. Figure 6.3 presents the relationship between wave height H, wave period T, water depth h and near seabed velocity amplitude U_w, according to first-order wave theory, g being the gravity constant.

The time variation can be assumed to be regular and described as a sine function, and the wave induced water particle acceleration can be derived from

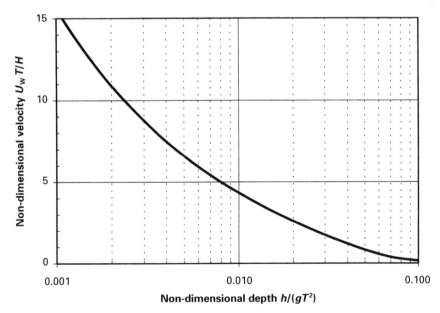

Figure 6.3 Diagram for calculation of wave induced seabed velocity amplitude

the velocity and the wave period at the seabed. The directionality of the waves (as well as the currents) has a significant influence on the on-bottom stability design. In connection with directionality it is useful to distinguish between two different phenomena:

- the short-term directional spreading of energy;
- long-term directionality.

Directional spreading of energy usually refers to short-term stationary wave conditions, in which the mean wave direction is constant. As the waves are short-crested the energy is distributed on different directions around the mean direction at any given time instant. Long-term directionality refers to the changes in mean direction with time, and is expressed as a long-term probability function. It should be emphasised that special effects, such as correlation or joint distributions of magnitude and mean direction, directional spreading and mean direction, etc. should be considered in connection with long-term directionality.

It is important to be aware of both types of directionality in the assessment of design data, so that the direction derived from the long-term statistical analysis that will lead to the highest velocity component perpendicular to the pipeline is identified, and so that the short-term distribution of energy around this mean direction is accounted for.

Spectral transfer of waves

Seastates may be described by a wave spectrum, for example by the distribution of wave energy on frequencies (and directions for three-dimensional waves). Wave spectra may be produced from wave measurements or they may be defined by an analytical expression, such as the Pierson–Moscovitz (PM) spectrum:

$$S_\eta(f) = \frac{5}{16} H_s^2 f_p^4 f^{-5} \exp\left[-\frac{5}{4}\left(\frac{f_p}{f}\right)^4\right]$$

where

$S_\eta(f)$	wave spectrum
f	wave frequency
H_s	significant wave height
f_p	peak frequency.

The above represents a spectrum for fully developed waves in deep water, i.e., no fetch or duration limitations. Each frequency component of the wave spectrum is transferred to the seabed analogous to the single wave transfer, see Figure 6.3. In the frequency domain the transfer function $H_{un}(f)$ is given by:

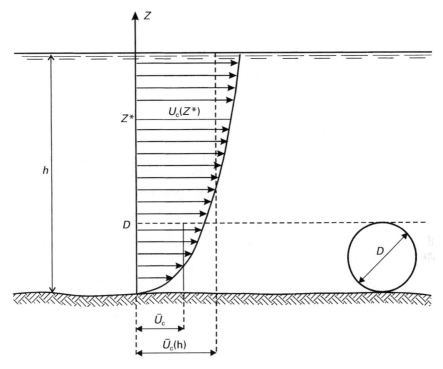

Figure 6.4 Definition of steady current velocity for different current profiles, $\bar{U}_c(h)$ depth average current, $\bar{U}_c$ average current over pipe diameter

$$H_{u\eta}(f) = 2\pi f/\sin h\,[k(f) \cdot h]$$

and the near bed velocity spectrum S_u is found as:

$$S_u(f) = H_{u\eta}^2(f)\,S_\eta(f)$$

The near seabed kinematics characterising the spectrum are as follows.
Significant velocity, U_{mo}:

$$U_{mo} = 2\sqrt{m_{ou}}$$

where

$$m_{ou} = \int S_u(f)\,\mathrm{d}f$$

(i.e. m_{ou} is the total energy of the velocity spectrum).
Maximum velocity amplitude, U_{max}:

$$U_{max} \cong 1.86\,U_{mo}$$

Peak period, T_p:

$$T_p \sim 1.4\,T_{02}$$

Mean zero-crossing period, T_{02}:

$$T_{02} = (m_{\mathrm{ou}}/m_{2\mathrm{u}})^{1/2}$$

$$m_{2\mathrm{u}} = \int f^2\,S_\mathrm{u}(f)\,\mathrm{d}f$$

T_{02} is a statistical measure that is approximately equal to the mean period T_z.

Steady current velocity

In the stability calculations the steady current averaged over the pipe diameter is used. The velocity profile $U_c(z)$ is described by the logarithmic expression:

$$U_c(z) = \frac{1}{\kappa} \cdot U_{\mathrm{fc}} \cdot \ln(30z/k_\mathrm{b})$$

where

κ	0.4 (von Karman's constant)
U_{fc}	shear velocity
k_b	seabed roughness
z	distance from seabed.

The mean current velocity over the pipe diameter (see Figure 6.4) is then:

$$\bar{U}_c = \frac{1}{D}\int_0^D U(z)\,\mathrm{d}z \approx \frac{1}{\kappa} \cdot U_{\mathrm{fc}}\ln\frac{30D}{2.72k_\mathrm{b}}$$

This velocity may easily be found for the following two situations:

(1) The velocity $U_c(z^*)$ is known at a height z^* above the seabed:

$$\bar{U}_c = U_c(z^*)\,\ln\!\left(\frac{30D}{2.72k_\mathrm{b}}\right)\!\bigg/\ln\!\left(\frac{30z^*}{k_\mathrm{b}}\right)$$

(2) The mean velocity over the entire depth, $\bar{U}_c(h)$, is known:

$$\bar{U}_c = \bar{U}_c(h)\,\ln\!\left(\frac{30D}{2.72k_\mathrm{b}}\right)\!\bigg/\ln\!\left(\frac{30h}{2.72k_\mathrm{b}}\right)$$

Current velocity profiles may be described using a 1/7th-power profile. Such a profile is generally in overall good agreement with the logarithmic profile. The drawback of using the power profile is that it is independent of the seabed roughness. In the case of an even seabed without significant roughness the 1/7th-power profile will underestimate the flow velocity near the seabed.

Seawater density
The density of the water at the level of the pipe is required as an input to the design calculations. The density depends upon the temperature, salinity and pressure, and is normally in the range 1000–1030 kg/m^3.

Sea level/water depth
The water depth is required in order to be able to calculate near seabed velocities.

The design water depth used in conjunction with the wave analysis should in general be the algebraic sum of the chart depth, corrected for the lowest astronomical tide (LAT), and the height of the storm surge corresponding to the design wave situation. However, the joint probability of the extreme wave height and the extreme storm surge is generally not known. Accordingly, a conservative approach would be to use the lowest storm surge (i.e. negative) combined with the lowest astronomical tide, which would result in the largest bottom velocities and thereby the largest hydrodynamic forces.

In principle the water depth associated with the design storm condition should be used in calculations. A safe approach is normally to use LAT in design, neglecting storm surge effects.

Hydrodynamic loads and soil resistance
The near seabed water velocity and acceleration are used to calculate the loads on the pipeline as described in Chapter 4. The soil resistance is treated in the succeeding Section 6.3.6.

Hydrodynamic stability check
The methods and procedures described in the following sections can be applied in on-bottom stability calculations in general, but are primarily for design against wave and current induced forces. Two basically different types of analyses can be used in the design, either a static analysis or a dynamic analysis.

The static analysis is applied when no movement of the pipeline is a requirement, and only two parameters are in principle necessary to characterise the wave induced water velocities:

- the velocity component perpendicular to the pipe axis;
- the period of the orbital velocity.

The time variation can be assumed to be regular, described as a sine-function, and the wave induced water particle acceleration perpendicular to the pipe axis can be derived from these parameters. The level at which flow conditions are calculated is the level of the pipe centre-line.

The static analysis is based on a two-dimensional quasi-static force balance between the hydrodynamic loads acting on the pipe and the soil resistance on the pipe. The result of the static analysis may be the required weight coating or other requirements to stabilise the pipeline.

The static analysis may be carried out as a significant static analysis or as a maximum load analysis. The significant static analysis applies significant wave parameters when calculating hydrodynamic loads, whereas the maximum load analysis applies the peak hydrodynamic loads in the stability check. The significant analysis implies that certain limited pipe displacements may occur in the design situation (e.g. less than 20 m), and the design methods should only be used when such displacements are acceptable. This will normally be the case if the seabed consists of loose sediments or soft clayey material. In case of hard seabed (rock or hard clay with boulders), damage of the pipeline or coating may occur in the case of pipe movement, and a maximum load design should be applied.

Dynamic analysis is applied when pipeline movements are accepted, and more detailed information on the near seabed flow conditions is needed in this situation. Information on the time variation of velocity and acceleration (time series), covering a significant part of a full storm is required. The minimum time span required should be sufficient to give a representative description of the peak of the storm. However, for certain applications longer time periods may be required, and parts of the rise and decay of the storm have to be included in the analysis. These time series will often be described in terms of a velocity spectrum, the velocities near the seabed for the dynamic analysis being determined by transferring surface wave data to the seabed using linear wave theory. The description of the surface waves may include three-dimensional spreading, i.e. the spreading of wave energy around the main direction of propagation.

The dynamic analysis involves a full dynamic simulation of a pipeline section resting on the seabed. The results of a dynamic analysis are the movements of the pipe and the pipe wall stresses. This analysis normally requires a sophisticated computer program. The DNV RP E305 *On-bottom stability design of submarine pipelines* presents a method that uses a set of non-dimensional parameters to calculate accumulated lateral displacements of a pipeline exposed to wave and current actions. The RP E305 procedures are presented in a graphical format, and are based on a large number of dynamic pipeline simulations. The American Gas Association project *Submarine pipeline on-bottom stability* has developed design procedures for on-bottom stability based on static as well as dynamic approaches.

6.3.3   *Design conditions and requirements*

The following design conditions should be analysed:

- the pipeline during installation;
- the pipeline during operation.

For each condition the stability analysis should be carried out for the most unfavourable pipe contents.

For the static design a safety factor of 1.10 is often adopted when assessing the stability. If a dynamic analysis is carried out, the lateral pipe displacements and pipe wall stresses should be checked against the allowable criteria. Depending on project requirements, and provided that stresses are within allowable limits, the maximum allowable lateral displacement can be:

- half the survey corridor;
- half the free distance to any fixed objects or other pipelines on the seabed;
- a fixed value, typically 20 m.

This lateral displacement criterion is only applicable if a dynamic analysis is carried out. Designs based on the significant static approach operate on a 'no displacement' concept, but implicitly accept exceeding the design criterion for the maximum wave in the design sea state.

The on-bottom stability assessment in the operational condition is normally performed for the 100-year design condition, i.e. the wave and current condition that has a mean return period of 100 years. If joint probability of waves and current is available, the 100-year wave and associated current or the 100-year current and associated wave should be used, whichever gives rise to the highest loads. Otherwise the following load combinations may be used as suggested by DNV RP E305.

- If waves dominate the hydrodynamic forces, the 100 year wave condition should be used in combination with the 10 year current condition.
- If current dominates the hydrodynamic forces, the 10 year wave condition should be used combined with the 100 year current condition.

For temporary phases (e.g. installation) the design condition has to be determined based on the period of exposure and on the risk involved and possible consequences of damage. A reasonable approach suggested by DNV RP E305 would be the following.

- For durations of less than 3 days, the environmental parameters can be established based on weather forecasts.
- For duration exceeding 3 days, a recurrence period of 1 year for waves and currents (for the relevant season) can be applied if there is no risk of loss of human life. If there is a risk of loss of human life, a recurrence period of

100 years (for the relevant season) should be applied – see above for the combination of waves and current. In no case should the season be taken to be less than two months.

If the temporary phase extends over one winter season, one or both of the following load combinations should be considered.

- If waves dominate the hydrodynamic forces, the 10 year wave condition should be used in combination with the 1 year current condition.
- If the current dominates the hydrodynamic forces, the 1 year wave condition combined with the 10 year current condition should be used.

6.3.4 *Static stability design format*

The static stability design is based on the following main assumptions.

- Pipe movements are not allowed, requiring equilibrium between loads (hydrodynamic forces) and reactions (soil resistance forces).
- Near bed wave flow is time varying and only the component perpendicular to the pipe axis is considered.
- Soil resistance is calculated based on two-dimensional assumptions, and may include simple friction as well as passive soil resistance.

The design format is expressed by:

$$F_H \leq \frac{1}{\gamma_s} R_H$$

where

$$R_H = \mu(W_s - F_L) + R_P$$

and

F_H in-line hydrodynamic force component (see Chapter 4)
R_H available horizontal soil reaction
μ coefficient of friction
W_s submerged weight
F_L lift force (see Chapter 4)
R_P passive soil resistance
γ_s safety factor against sliding.

The design format may be expressed as a requirement to submerged weight:

$$W_s \geq \frac{\gamma_s}{\mu}(F_H - R_P) + F_L$$

The recommended factor of safety is $\gamma_s = 1.10$.

6.3.5 *Dynamic stability design format*

The dynamic stability design is based on the following main assumptions.

- Pipe movement is allowed, and restrictions on total movements or maximum stresses form the design criteria.
- The pipeline is viewed as a structural unit (long section used in the analysis), where bending and tensile stresses act as restoring forces.
- The wave flow is modelled as three-dimensional with a mean direction and energy spreading.
- The effect of pipe movements on the hydrodynamic forces is included.

The design format is either expressed through a maximum allowable lateral displacement or by the general stress criteria for the pipeline design.

Axial loads in heated pipelines need special attention because of the potential risk of lateral buckling including large displacements and high bending moments, see Section 6.5.

6.3.6 *Pipe–soil interaction*

The pipeline configuration on the seabed, during installation as well as in the operational phase, is largely determined by the interaction between the pipeline and the seabed soil. Pipe–soil interaction is important for the stability of the pipeline on the seabed in the horizontal, as well as in the vertical, direction. Furthermore, pipe–soil interaction plays an important role in free span assessment and in pipeline expansion and buckling, see Sections 6.4 and 6.5. The major parameters are discussed below.

Axial force

The axial force may have a dominant influence on the deflections in areas where transversal pipeline movements occur. The following components contribute to the axial force in a pipeline on the seabed:

- residual force from installation, N_{res};
- thermally induced axial force, N_θ;
- pressure induced axial force, N_p;
- hoop stress induced axial force, N_v;
- displacement induced axial force (non-linear), N_{nl}.

The various axial force components are considered positive as tension, and are evaluated as follows.

- The residual axial force evaluated from construction activities is difficult to determine accurately. Normally it is a conservative approach to neglect residual forces from the installation (e.g. lay tension). Exceptions are formed by cases where the residual tension is a dominant parameter determining the pipeline configuration, such as free spans (see Section 6.4) or horizontal curves (see below).
- The temperature induced axial force occurs if the pipeline is operating at temperatures that are different from when it was installed. The force component resulting from the prevented temperature deformations can be calculated as:

$$N_\theta = -\pi \ (D_s - t)\, t\, E\, \alpha\, \Delta\,\theta$$

where

α	steel temperature expansion coefficient (in $°C^{-1}$)
$\Delta\theta$	temperature increase (in $°C$)
D_s	steel pipe outer diameter
t	steel pipe wall thickness
E	steel modulus of elasticity.

The temperature at installation will be that of the environment for operating uninsulated pipelines a long distance from the inlet, and the temperature of the medium will be close to ambient, in which case the temperature force component N_θ can be neglected. This is so for transmission pipelines, whereas for flowlines the temperature will often be the dominant contributor to the axial force.

- The pressure induced axial force is the difference between the internal pressure acting on the pipe bore and the external pressure acting on the total pipeline cross-section:

$$N_p = -\frac{1}{4}\pi(D_s - 2t)^2\, p_i + \frac{1}{4}\pi D_s^2\, p_e$$

where

p_i	internal pressure
p_e	external pressure.

Note that this component does not affect the pipe wall, but is a force exerted by the pipe medium, balanced by soil friction on the pipeline. At a free end, where the friction vanishes, it becomes an end cap force, giving rise to tension – and corresponding elongation – in the linepipe steel. For pipelines provided with an impermeable external coating the steel outer diameter D_s in the last (external pressure) term should be replaced by the coated pipe outer diameter.

- The hoop stress induced axial force results from the prevented Poisson contraction, which is proportional to the hoop stress:

$$N_v = \frac{1}{2}\pi(D_s - 2t)(D_s - t)\,\nu\,(p_i - p_e)$$

where

ν steel Poisson's ratio.

- The displacement induced axial force is generated by the deflection of the pipeline. This is a priori unknown, but in the case of a free span it can be found through an iterative scheme of free span analysis.

To be totally consistent, the treatment of the internal pressure should distinguish between the static pressure, which gives rise to stresses in the pipe wall, and the total pressure on the pipe bore, which includes the velocity term p_v:

$$p_v = \rho_i v^2$$

where

ρ_i density of the transported medium
v flow velocity.

As it is the total pressure that is represented by the design pressure, the velocity pressure will lower the static pressure, reducing the Poisson term. For steel pipelines this will be normally be insignificant compared with the temperature term, but it can have an effect for flexible pipelines, where the pressure terms dominate the axial force.

The total axial force N_a (positive as tension) of a fully restrained pipeline is made up by the above contributions:

$$N_a = N_{res} + N_\theta + N_p + N_v + N_{nl}$$

It is useful to distinguish between the 'true wall' axial force and the 'effective' axial force. The effective axial force, which governs the global behaviour of the pipeline, may be written as:

$$N_a = N_w - p_i A_i + p_e A_e$$

where

N_w axial force in the pipe wall
A_i internal cross-sectional area of the pipe (bore)
A_e total cross-sectional area of the pipe (including coating).

Thus, a pipeline may be subject to global buckling due to internal pressure even if the pipe wall is in tension.

Vertical bearing capacity

The vertical bearing capacity is specifically of concern in cases where very heavy pipelines or sediments that are soft or subject to liquefaction may lead to uncontrolled sinking of the pipe. Otherwise, the vertical bearing capacity may be used to establish the soil reaction displacement relation, which can be used when calculating the equilibrium configuration of a free span. The vertical bearing capacity of the soil may be found using the generalised formulation for the ultimate bearing capacity of a strip footing.

If the seabed is very soft the pipe will sink into the seafloor until the soil reaction can balance the downward forces. The stability of the pipeline (in the vertical direction) is checked by establishing the force equilibrium of all relevant forces, Figure 6.5 illustrates the force equilibrium.

$$W_S + F_V + F_S - 2F_f \le R_V$$

where

W_S submerged weight
F_V vertical force due to pipe curvature in the vertical plane
F_S weight of soil on top of pipe

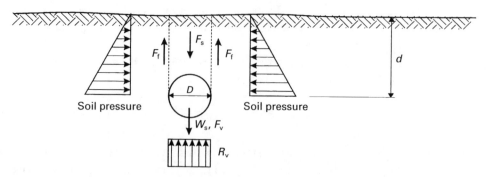

Figure 6.5   **Vertical equilibrium for a buried pipe**

F_f friction along shear planes
R_V Bearing capacity of the soil.

The force components are calculated as:

$F_V = N_a/r$ (axial force divided by radius of curvature)

$$F_f = \frac{1}{2}\gamma_s d^2 \tan\varphi_s \quad (F_f \le \frac{1}{2}F_s)$$

$$R_V = \left(\frac{1}{2}\gamma_s D N_\gamma + \gamma_s d N_q + C_u N_c\right)D$$

where

N_γ, N_q, N_c bearing capacity factors.

Lateral soil resistance

The main concern is to analyse the ability of the soil to resist lateral forces imposed by the pipeline. Owing to the large torsional stiffness of steel pipelines, only the soil resistance against sliding is of concern. Rigorous analytical solutions are not available for prediction of lateral resistance of pipes in sand or clay. However, a general approach has been established by separating the total soil resistance into two parts – one which is purely frictional, and one which is due to a passive soil pressure. The passive soil pressure is generated by the pipe embedment, the cyclic loads from wave action, or by the build-up of a soil ridge in front of a sliding pipe.

In sand the lateral soil resistance is separated into a Coulomb frictional component and a passive soil resistance component. The frictional component depends on the vertical reaction force and the coefficient of friction. The passive component depends on the submerged unit weight of the sand, the height of the sand ridge in front of the pipe, the pipe diameter, and the loading conditions. An approximate expression for calculating the lateral soil resistance is given below:

$$R_H = \mu R_V + \frac{2}{3}\beta \rho_s D^2 (H_u/D)^{3/2} \quad 0 \le H_u \le D$$

where

R_H lateral soil resistance
μ pipe–soil coefficient of friction
R_V vertical reaction force
ρ_s submerged unit weight of sand
D external diameter of (coated) pipe
H_u height of soil ridge
β empirical constant.

Table 6.10 Friction coefficients for calculating lateral soil resistance (values are given for guidance only)

	Lateral friction coefficient, μ			
Coating	Concrete		Steel, Epoxy	
Analysis	Static	Dynamic	Static	Dynamic
Loose sand	0.7	0.5	0.5	0.4
Dense sand	0.7	0.5	0.5	0.4
Silty sand	0.6	0.4	0.5	0.3
Soft clay	0.4	0.2	0.4	0.2
Medium clay	0.4	0.2	0.3	0.2
Hard clay	0.5	0.2	0.4	0.2
Rock	0.5	0.4	0.4	0.4

Table 6.11 Values of β for sand and clay. Experimental justification of β is only available for medium dense sand and medium clay (values are given for guidance only)

Soil type	Sand	Clay
Soil reaction constant β	10 (3–15)	2 (1–3)

In clay the same empirical approach as followed for pipelines on sandy soils can be used, except that the passive soil resistance depends on the undrained shear strength:

$$R_{\mathrm{H}} = \mu R_{\mathrm{V}} + \beta C_{\mathrm{u}} D(H_{\mathrm{u}}/D) \quad 0 \le H_{\mathrm{u}} \le D$$

where

C_{u} undrained shear strength of clay.

The remaining parameters are defined above.

Appropriate values for the pipe–soil friction coefficient μ and the empirical parameter β are given in Tables 6.10 and 6.11.

Horizontal curves

Pipeline installation along a prescribed route would generally include sections that are curved in the horizontal plane. The radius of curvature is normally so large that the bending resistance of the pipe can be neglected, but tension induced in the pipeline during laying will tend to straighten any curves, unless the pipe is kept in place by the lateral soil resistance. The minimum lay radius that can be achieved can be calculated from:

$$r_{\min} = \gamma N_{\mathrm{res}} / R_{\mathrm{H}}$$

where

$r_{\min}$	minimum lay radius
γ	safety factor
N_{res}	residual lay tension
R_{H}	lateral soil resistance.

The factor of safety might be set at $\gamma = 1.2$, to account for variations in tension setting and dynamical effects, and the lateral resistance is calculated from the expressions above.

Axial soil resistance

The methodology for calculating the axial soil resistance on buried or partly buried pipeline sections depends on whether the soil is classified as sand or clay. The axial soil resistance is used when calculating the effective axial force in the free span. The axial friction restrains the deflection induced axial movement at the supports, and thereby limits the deflection and bending moments.

In sand, the axial soil resistance may be found using a friction coefficient and the soil pressure perpendicular to the pipe surface:

$$R_{\mathrm{a}} = \int_A \mu_{\mathrm{a}} \sigma_{\mathrm{n}} \, \mathrm{d}A$$

where

$$\mu_{\mathrm{a}} = \tan(f_\phi \, \phi_{\mathrm{s}})$$

In clay a similar expression is valid, but it is either the undrained shear strength or the effective friction angle (drained failure) that determines the axial resistance. The minimum resistance found by the following expressions should be used:

$$R_{\mathrm{a}} = \int_A \mu_{\mathrm{a}} \sigma_{\mathrm{n}} \, \mathrm{d}A \quad \text{and} \quad R_{\mathrm{a}} = \int_A f_{\mathrm{c}} C_{\mathrm{u}} \, \mathrm{d}A$$

The integral is taken over the surface area A in contact with the soil, and the parameters are:

R_{a}	axial soil resistance
μ_{a}	axial friction coefficient
σ_{n}	normal pressure on the soil
f_ϕ	skin friction factor (sand or clay)

Table 6.12 Skin friction factors to be applied when calculating the axial soil resistance

Pipe surface condition	Material	Sand	Clay	
		f_ϕ	f_ϕ	f_c
Smooth	Steel, plastic	0.60	0.50	0.25
Medium rough	Rusted steel	0.80	0.50	0.45
Rough	Concrete	0.92	0.70	0.50

ϕ_s effective angle of friction (sand or clay), see Section 3.2.1
f_c skin friction factor (clay).

Table 6.12 gives typical values for the factors. The effective angle of friction and the undrained shear strength should be determined by standard geotechnical tests.

6.4 Free span evaluation

6.4.1 *General*

The pipeline free span shall have adequate safety against the following failure modes and deformations:

- excessive yielding;
- fatigue;
- buckling;
- ovalisation.

Free span analysis should be based on generally accepted static and dynamic calculation methods, including non-linear structural modelling, soil reaction description, and deflection induced axial forces. The following pipeline conditions should be considered:

- empty pipeline;
- water-filled pipeline;
- pipeline during hydrotesting;
- operating pipeline.

The analysis of free spans normally requires:

- static analysis for determining pipeline configuration, sectional forces, and stresses under functional loads;
- eigenvalue analysis for determining natural frequencies and modal shapes;

- dynamic analysis for determining pipeline deflection, sectional forces, and stresses under combined functional and environmental loads or accidental loading;
- fatigue analysis for determining accumulated fatigue damage due to cyclic loads from wave action and vortex shedding.

In addition, free spans must be evaluated with respect to trawl pull-over and hooking, i.e. the situation where trawl equipment is wedged under the pipeline, see Section 6.2.6.

The methodologies and design cases considered are well suited for Load and Resistance Factor Design (LRFD) as adopted by the Recommended Practice DNV RP F105 *Free spanning pipelines*, for example, which includes a detailed description of methods for analysis of pipeline free spans exposed to wave and current action. The methods and procedures for pipeline free span assessment introduced in the sub-sections below present overall general requirements to free span design. The requirements are based on maximum allowable stress or fatigue damage and usage factors that assure that the required safety is present.

6.4.2 *Free span classification*

Free spans may be classified as isolated or interacting. Two spans are isolated if the intermediate support length is such that the static and dynamic behaviour of each span is unaffected by the presence of the other. In all other cases, free spans are interacting. Furthermore, in order to obtain a realistic application of load in free span evaluation it is necessary to distinguish between free spans that are caused by irregularities of the seabed and free spans that develop after pipeline installation due to some scouring action on the seabed, see Table 6.13.

Unevenness induced free spans are free spans that have existed since the installation, caused by irregularities of the seabed, which change only marginally in length during the pipeline lifetime (excluding the effect of intervention works). In this case, the as-laid pipeline configuration is to be determined prior to applying the remaining functional loads.

Scour induced free spans are generated by scour or other seabed instabilities, and may change in position and length throughout the pipeline lifetime. The

Table 6.13 **Free span classification**

Free span classification	Implication
Isolated	Simple structural model
Interacting	Complex structural model
Unevenness induced	Loading sequence in steps
Scour induced	All loads applied in one step; span may change in time

evaluation of the possibility of erosion and seabed instabilities is usually a specific project activity aiming at determining the maximum expected free span lengths and exposure periods, while the actual location of the free span in most cases is unpredictable.

6.4.3 *Pipeline and free span data*

The pipeline data are required for the determination of the loads and resistance parameters and the structural static and dynamic properties of the pipeline free span. The pipeline data may be available from the original design documents, but up-to-date information or on-site inspection data may be required to confirm the actual span configuration, and that no damage to the pipeline has taken place.

Pipe data
The following pipe data are required:

- outer diameter of steel pipe, D;
- wall thickness, t;
- thickness and density of any corrosion coating;
- thickness and density of any weight coating;
- modulus of elasticity;
- Poisson's ratio of steel;
- steel yield stress;
- structural damping parameter;
- coefficient of temperature expansion;
- pipe roughness including marine growth;
- pipe orientation on the seabed.

Functional data
The following functional data are required:

- density of contents at operating pressure;
- excess temperature of contents in operating conditions;
- internal/external pressures;
- residual forces from installation operations.

Free span data
The following information is required:

- water depth;
- seabed roughness;
- free span classification (isolated or interacting, unevenness or scour induced);

- free span profile;
- free span length;
- pipe clearance above seabed;
- pipe embedment on shoulders;
- soil classification and data.

Hydrographic data

Pipeline free span assessment requires input in the form of:

- wave induced (orbital) water velocities and accelerations at the level of the pipeline span;
- steady current velocity at the pipeline span;
- seawater density, ρ.

In order to perform the analyses, the wave and current information has to cover a representative range of conditions. This is normally fulfilled when wave data are specified in the form of a scatter diagram in significant wave height, H_s, and mean period, T_{02}, covering a period of one year. Current data should preferably be specified as current velocity and direction associated with each seastate defined in the wave scatter diagram. Alternatively, current may be specified in a scatter diagram by current velocity and direction.

Geotechnical data

The evaluation of free spans requires information on the geotechnical conditions at the free span location. Representative soil data and parameters are required when evaluating the support conditions and the damping properties of the free span. Furthermore, the erodibility of the seabed material should be evaluated when assessing the possible development in the free span configuration. Section 3.2.1 describes the recommended soil classification, and gives typical values for relevant soil parameters. The parameters should preferably be obtained by means of standard geotechnical tests, performed on undisturbed soil samples taken in the area of interest.

6.4.4 *Static analysis*

The static analysis includes only the functional loads that may give rise to insignificant dynamic amplification of the response. The analysis should as a minimum include the effect of the following phenomena and conditions:

- soil–pipe interaction;
- non-linear relationship between lateral deflection and axial force;
- correct sequence of loading;
- presence of adjacent spans when interacting.

The loading sequence depends on whether the span is classified as scour induced or unevenness induced.

- In the case of a scour induced free span, the equilibrium configuration has to be determined under application of all the loads, starting from a rectilinear configuration.
- In the case of an unevenness induced free span, an intermediate equilibrium configuration (as-laid configuration) has to be determined using loads corresponding to empty pipe conditions, a constant axial force equal to the laying tension (effective), and no axial restraint. The final equilibrium configuration is to be determined starting from the intermediate one, under application of the remaining loads and the actual axial restraint.

In the flexural analysis the bending stresses are superimposed upon axial stresses from the forces discussed in Section 6.3.6 above. Verification of the linepipe design may be carried out in terms of the von Mises stress, assuming a state of plane stress in the pipe wall:

$$\sigma_e = (\sigma_1^2 + \sigma_2^2 - \sigma_1 \sigma_2)^{1/2} \leq f_y$$

where

σ_e equivalent von Mises stress
σ_1, σ_2 principal stresses
f_y design steel yield stress.

For the temporary situation, during installation of the pipeline, the following cases are considered:

- empty pipeline;
- waterfilled pipeline;
- waterfilled pipeline under hydrotest pressure.

Obviously the gravity loads are highest for the waterfilled pipeline, but the added submerged weight may cause additional seabed contact, thus eliminating or reducing some free spans.

    The steel wall of an operating pipeline is subjected to a hoop stress σ_H from the internal pressure, and an axial stress σ_A from the pressure and temperature induced axial pipeline restraint force discussed above. If the pressurised pipe is subject to additional bending, the bending stresses $\pm\sigma_B$ are superimposed upon these stresses, thus in the absence of torsion the extreme principal stresses are, see Figure 6.6:

$$\sigma_1 = \sigma_A \pm \sigma_B \quad \text{and} \quad \sigma_2 = \sigma_H$$

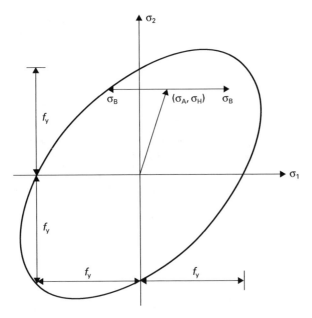

Figure 6.6   von Mises yield locus with stress components

In the figure, the flexural capacity σ_B can be read off as the horizontal distance to the closest point on the yield locus from the end of the stress vector $(\sigma_1, \sigma_2) = (\sigma_A, \sigma_H)$, and if the pipeline is highly pressurised there is virtually no elastic bending capacity left. It is obvious that the maximum capacity (simultaneous yield at both faces) is obtained for $\sigma_A = \sigma_H/2$, but normally the axial tension is less than half the hoop stress, and yielding will start at the compressed face. This yielding will produce permanent strains, which upon depressurisation will give rise to residual tensile stresses. When the pipeline is pressurised again, the compressive bending stresses will first have to overcome this residual tension, leading to simultaneous yielding at the compression and the tension face. This phenomenon is known as elastic shakedown, the shakedown stress distribution being the one corresponding to $\sigma_A = \sigma_H/2$.

Thus by allowing one-time yielding and taking account of elastic shakedown (Braestrup 1993) the analysis can be based upon simultaneous yielding at the tension and compression faces, irrespective of the actual axial stress. It might even be argued that by local plastic deformations during hydrostatic testing the pipeline could adapt to seabed irregularities, However, the calculations are carried out with design values and, due to the safety factors involved, actual plastic deformations are highly unlikely.

If a computer program based on hoop stress alone (lay tension and temperature neglected) is used, the elastic shakedown can be simulated by adopting the value $\nu = 0.5$ for Poisson's ratio (Braestrup 1989). Similar considerations apply to the analysis of expansion offsets; see Section 6.5.2 below.

Note that many codes do not allow for elastic shakedown, but regard first yield as failure. DNV OS-F101 specifies a limiting interaction between the imposed axial force, bending moment and internal overpressure.

6.4.5 *Dynamic analysis*

The pipeline free span is a dynamic structure, having well defined natural frequencies and modes. Such a structure is susceptible to amplified response when exposed to cyclic loads having frequency close to the natural frequency. Therefore, free span analysis shall include wave load analysis, accounting for dynamic amplification and vortex shedding. The dynamic analysis includes all loads that may give rise to significant dynamic amplification of the response. The loads that should be treated in a dynamic analysis are: loads from wave action, vortex shedding, and impact loads.

The dynamic analysis requires an eigenvalue analysis of the free span for determination of natural frequencies and modal shapes. As the eigenvalue analysis is a linear analysis a consistent linearisation of the problem must be made. The analysis should account for the static equilibrium configuration, and the linearised stiffness of the soil shall take into account the correct properties of the soil. Special attention must be paid to the definition of the axial stiffness of the soil, as the results of the eigenvalue analysis in the vertical plane are very much affected by this axial stiffness. Where only the suspended span is modelled, the boundary conditions imposed at the ends of the pipeline section should represent the correct pipe–soil interaction and the continuity of the entire pipe length. The influence of added mass as a function of seabed clearance has to be considered when calculating natural frequencies.

The damping K_{si} of a free span is one of the parameters determining the maximum response to hydrodynamic loads. The damping is expressed by the stability parameter for each natural mode or eigenvector:

$$K_{si} = \frac{4\pi m_{ei} \zeta_{Ti}}{\rho \cdot D^2}$$

where

ζ_{Ti} total damping ratio
m_{ei} effective mass per unit length of the pipe
i index relating to the ith mode.

The remaining parameters are defined in Section 6.4.3.

The total structural damping ratio, ζ_{Ti}, includes:

- structural damping, ζ_{str}, due to internal friction forces of the pipe material, depending on strain level and associated deflections;

- soil damping, ζ_{soil}, (internal and radiation damping) due to pipe–soil interaction at the supports;
- hydrodynamic damping, ζ_h, accounting for the effects of the non-uniformity of the flow velocity and variation in the pipe–seabed clearance.

The structural damping is generally expressed as a damping ratio. If no information is available the structural damping ratio $\zeta_{str} = 0.005$ can be assumed. If the pipeline is coated with concrete the sliding at the interface between the concrete and the corrosion coating can further increase this value.

The contribution of the soil to the overall damping ratio of the pipe–soil system can be significant, and should be evaluated.

The effective mass should include contributions from structural mass, hydrodynamic added mass and mass of the content.

6.4.6 *Fatigue analysis*

Dynamic loads from wave action, vortex shedding, etc. may give rise to cyclic stresses, which may cause fatigue damage to the pipe wall, and ultimately lead to failure. The fatigue analysis should cover a period that is representative for the free span exposure period, and fatigue calculations should only be applied to the pipeline conditions that are of such duration that noticeable damage may occur. Fatigue calculations are therefore normally neglected for the hydrotesting conditions.

The fatigue damage from vortex induced vibrations (VIV) should be calculated, including as a minimum:

- dynamic effects when determining stress ranges;
- calculation of the number of cycles in a representative number of stress ranges;
- calculation of fatigue damage according to the Palmgren–Miner accumulation law;
- determination of the number of cycles to failure using a suitable S–N curve.

The stress ranges to be used in the fatigue analysis may be found using two different methods.

(1) The stress ranges are found by a dynamic analysis applying an external load to the free span (load model).
(2) The stress ranges are determined using the normalised response amplitudes for a given flow situation (response model), appropriately scaled to the real free span.

Both methods may be applied to a wide range of flow conditions, and the use of one particular method is primarily determined by practical reasons or by the quality of the appropriate model for the actual case. Appropriate response models may be found in DNV RP F105 *Free spanning pipelines*, which recommends that the following flow conditions be considered:

- cross-flow VIV in steady current and combined wave and current;
- in-line motion due to cross-flow VIV;
- in-line VIV in steady current and current dominated flow.

S–N curves

The number of cycles to failure is defined by an S–N curve of the form:

$$N = C\,(S)^{-m}$$

where

N	number of cycles to failure at stress range S
S	stress range based on peak-to-peak response amplitudes
m	fatigue exponent (the inverse slope of the S–N curve)
C	characteristic fatigue strength constant defined as the mean-minus-two-standard-deviation curve (in $(MPa)^m$).

The constants m and C may change for a given S–N curve when the number of cycles exceeds a certain threshold value, typically 10^6 or 10^7.

The S–N curves (material constants m and C) may be determined from:

- dedicated laboratory test data;
- fracture mechanics theory;
- accepted literature references.

If detailed information is not available, the S–N curves given in DNV RP C203 *Fatigue strength analysis of offshore steel structures* may be used, corresponding to cathodically protected carbon steel pipelines.

The S–N curves may be determined from a fracture mechanics approach, using an accepted crack growth model with an adequate (presumably conservative) initial defect hypothesis and relevant stress state in the girth welds. Considerations should be given to the applied welding and NDT specifications.

A stress concentration factor (SCF) may be defined as the ratio of hot spot stress range over nominal stress range. The hot spot stress is to be used together with the nominated S–N curve.

Stress concentrations in pipelines are due to eccentricities resulting from different sources. These may be classified as:

- concentricity, i.e. difference in pipe diameters;
- difference in thickness of joined pipes;
- out-of-roundness and centre eccentricity.

The resulting eccentricity, δ, defined as the offset of the centre plane in the pipe wall across the weld, may conservatively be evaluated by a direct summation of the contribution from the different sources. In the absence of data the DNV RP C203 suggests taking $\delta = 0.15t$, with a maximum of 4 mm. The eccentricity δ should be accounted for in the calculation of the stress concentration factor, SCF, and the following formula is proposed by DNV RP C203:

$$SCF = 1 + \frac{3\,\delta}{t}\exp[-(D/t)^{-0.5}]$$

The parameters are defined in Section 6.4.3.

Fatigue damage

The fatigue damage may be assessed on the basis of the Palmgren–Miner accumulation rule. This implies replacing the actual stress range distribution by a histogram with a number, I, of constant amplitude stress range blocks, with corresponding stress ranges S_i. The fatigue damage is then calculated as:

$$D_{fat} = \Sigma \frac{m_i}{N_i}.$$

where

D_{fat}	accumulated fatigue damage
m_i	number of stress cycles with stress range, S_i
N_i	number of cycles to failure at stress range, S_i (defined by the S–N curve).

The summation is in principle performed over all stress cycles in the design life, and the stress cycles S_i (number and magnitude) may be calculated using a load model, through integration of the equation of motion, or through the application of a response model.

Safety factor

The allowable fatigue damage ratio depends on the safety class, and the values recommended by DNV RP F105 are stated in Table 6.14.

It should be noted that these factors are applied together with other partial safety factors in DNV RP F105.

The accumulated fatigue damage of the pipeline is incurred during the following phases:

- installation (typically pipelaying);
- on the seabed (empty and/or waterfilled);
- operation.

Table 6.14 Allowable damage ratio for fatigue

Safety class	Low	Normal	High
α_{fat}	1.0	0.5	0.25

To ensure a reasonable fatigue life in the operational phase it is common industry practice to assign no more than 10% of the allowable damage ratio to the two temporary phases (as mentioned above, hydrotesting is normally neglected).

6.5 Expansion and global buckling

6.5.1 *General*

If a pipeline were free to elongate, the rise in temperature and pressure during operation would result in an increase in length. However, due to the restraint offered by the seabed friction, such pipeline expansions only manifest themselves at the ends, i.e. at the tie-in points to fixed structures. At undisturbed sections of the pipeline the restraint against thermal and pressure induced expansion may cause a compressive pipeline force, which could result in a global buckling mechanism.

Realistic analysis methods and design criteria for these phenomena are fundamental for establishing practical installation specifications and to minimise costly mitigation measures, principally rock dumping.

Referring to the axial force components introduced in Section 6.3.6, and neglecting any variations in residual forces from installation (lay tension), the total compressive force N_0 in a restrained pipeline due to the increases in temperature and pressure is found as:

$$N_0 = \pi (D_s - t)\, t\, E\, \alpha\, \Delta\theta + \frac{1}{4} \pi (D_s - 2t)^2\, \Delta p - \frac{1}{2} \pi (D_s - 2t)(D_s - t)\, v\, \Delta p$$

$$= E\, A\,[\alpha\, \Delta\theta - \varepsilon_a(\Delta p)] + A_i\, \Delta p$$

where

$\quad A \qquad$ cross-sectional area of the pipe wall

$\quad \varepsilon_a(\Delta p) \quad$ axial pipe wall strain induced by the internal pressure increase Δp.

The remaining parameters are defined in Section 6.3.6.

The effective axial force N_0 (positive as compression) is the sum of a compressive force in the pipe wall (which is the difference between the compression induced by the restrained temperature expansion and the tension induced by the restrained Poisson contraction) and a compressive force in the pressurised pipe medium.

Pipeline expansion is only dependent upon the above force due to variations in temperature and pressure. An assessment of the resistance to buckling, however, requires consideration of the total effective axial force evaluated in Section 6.3.6.

6.5.2 *Pipeline expansion*

For undisturbed pipeline sections the compressive force (indicated by the symbol $\ominus$ in Figure 6.7) induced by temperature and pressure is balanced by axial friction forces from the surrounding soil. If the axial component of the friction restraint disappears at some point, either because the pipeline emerges from the seabed or a 90° bend is introduced, the pipeline would remain stationary only if a restraint force $N = -N_0$ were applied at the end point, see Figure 6.7a. Removal of this imaginary restraint force results in an expansion of the pipeline, and the

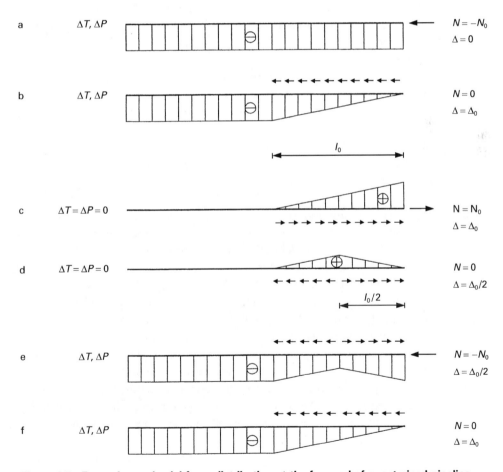

Figure 6.7   Expansion and axial force distribution at the free end of a restrained pipeline

expansion $\Delta = \Delta_0$ is determined by the requirement that the friction force mobilised over the expansion length, l_0, shall equal the compressive force, N_0. The resulting pipeline force distribution is shown in Figure 6.7b, under the assumption that the friction is uniformly distributed over the expansion length.

Now suppose that the pipeline is cooled and depressurised to the conditions at installation, i.e. $\Delta\theta = p_i = 0$. Then keeping the pipeline fixed would require a tensile force $N = N_0$ at the pipeline end and, since the compressive pipeline force vanishes, the pipeline force distribution would be as shown in Figure 6.7c, the net tension being reduced to zero over the length l_0. Removal of the imaginary end tension results in contraction of the pipeline, the amount being determined by the requirement that the mobilised friction shall balance the pipeline force at the end of the retraction length. As seen from Figure 6.7d, this implies that the retraction is $\Delta_0/2$ over the length $l_0/2$, (Palmer and Ling 1981).

The result of the complete cycle is that the pipeline has undergone a permanent expansion displacement $\Delta = \Delta_0/2$, and that a residual tension balanced by seabed friction has been introduced in the pipeline. The residual tension is triangularly distributed, with a maximum $N = N_0/2$ at the distance $l_0/2$ from the pipeline end.

When the pipeline is again brought into operation, the compressive pipeline force returns, and is imagined to be balanced by an end restraint $N = N_0$. However, because of the residual tension in the pipeline the compressive force is reduced near the end, see Figure 6.7e. Thus removal of the imaginary end restraint force results in a reversal of the friction on the extreme length $l_0/2$ only, and the pipeline expands by $\Delta_0/2$ to the total expansion $\Delta = \Delta_0$, see Figure 6.7f. The state of the pipeline is now identical to the situation after the first heating and pressurising.

The above considerations imply that although the first operation of the pipeline results in an free end expansion $\Delta = \Delta_0$, repeated shut-down/start-up cycles will only lead to pipeline expansions of $\Delta_0/2$, the total expansion varying between $\Delta = \Delta_0$ and $\Delta = \Delta_0/2$. Thus the variable imposed deformation that shall be accommodated at a pipeline end is $\Delta = \Delta_0/2$.

It has been speculated (Driskill 1981) that time-dependence of soil properties might lead to accumulating pipeline expansions as a result of operating cycles. Indeed, it is likely that soil creep will imply a lower friction factor for the long term operation periods than for the relatively brief shut-down periods. The traditional way of dealing with this is to assume conservative values for the skin friction factors, see Section 6.3.6, resulting in an overestimation of the initial expansion.

Also creep of the corrosion coating and straightening of the pipeline on the seabed (snake effect) have been cited as reasons for accumulating pipeline expansions at platforms. However, it appears that only minor contributions can be ascribed to these effects (Loeken 1980).

The pipeline expansion at an offshore platform or subsea wellhead may be so small that it can be absorbed by the structure, for example, by bending of the riser. In many cases, however, it is necessary to introduce a horizontal bend on the seabed, in the form of a U, Z- or L-spool between the end of the pipeline and the foot of the riser. The statical system and deformations are visualised schematically in Figure 6.8, a 90° bend being assumed for simplicity. The pipeline and the spool are assumed to be fixed against rotations at the distances h and l, respectively, from the bend. The expansion of the pipeline is termed Δ, and the resistance against lateral displacement (friction and passive soil pressure) is modelled by the uniformly distributed load p. A more accurate analysis would take account of the elastic soil springs, the expansion of the spool, and the lateral displacement of the pipeline.

The resulting moment distribution is shown in Figure 6.8c, where the redundants are termed M_1, M_0 and M_2. With increasing values of Δ, yielding will start at one of four locations:

- at the pipeline clamped section (M_1);
- at the bend (M_0);
- at the spool maximum moment (M_x);
- at the spool clamped section (M_2).

The location and magnitude of the maximum moment M_x depends upon the values of p, l, M_0 and M_2. The detailed analysis will determine which location is critical, but experience shows that yielding is most likely to commence at the bend. The corresponding yield moment is termed M_Y, and it is assumed to occur for a pipeline expansion $\Delta = \Delta_Y$.

Referring to Figure 6.8, it is obvious that the magnitude of the negative (closing) moment M_0 produced by the pipeline expansion can be reduced if a positive (opening) moment is introduced at the bend before the pipeline is put into operation. This can be achieved by cold-springing the spool against the pipeline during tie-in; see Figure 6.9, a practice which has been used in the North Sea.

The spool and the pipeline are aligned on the seabed such that a gap Δ_c is left between the bend and pipeline flanges. Closing of the gap by moving the spool flange, the pipeline remaining fixed, will produce an opening moment in the bend, and by choosing $\Delta_c = \Delta_Y$ incipient positive yielding is introduced. This means that a temperature expansion of $\Delta = 2\Delta_Y$ can be accommodated before negative yielding occurs in the bend.

If no cold spring is introduced then temperature expansions $\Delta > \Delta_Y$ will produce negative yielding in the bend. If the pipeline cools down again, the contraction will result in a positive moment, analogous to the one produced by cold-springing. A new temperature expansion will have to overcome this residual moment before producing negative bending. Thus the effect is identical to that of the cold spring, except that a one-time yielding has occurred, whereas elastic conditions are maintained throughout by cold-springing. The maximum effect is obtained by analytically introducing a residual moment $M_0 = M_Y$, corresponding to incipient

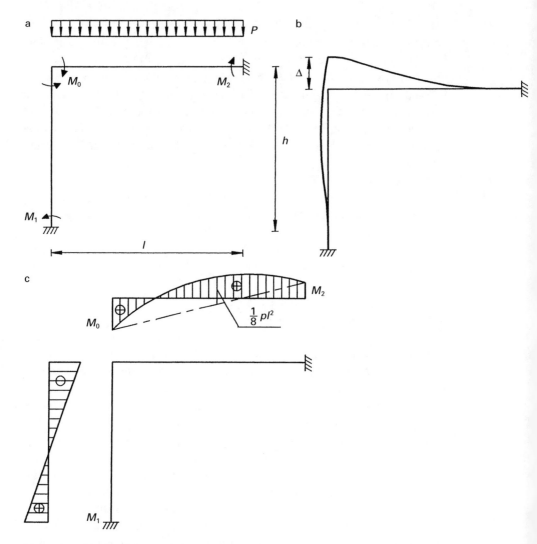

Figure 6.8 Pipeline bend analysis

yielding at the bend. Then a temperature expansion of $\Delta = 2\Delta_Y$ can be accommodated without repeated yielding.

The phenomenon is analogous to the elastic shakedown of the cross-sectional steel stresses in a pipeline subjected to simultaneous bending and pressurisation, discussed in Section 6.4.4. In combination with the expansion analysis considered above, the result is that the initial pipeline expansion that can be tolerated is four times the expansion that produces yielding in bending of the pipeline (Braestrup 1993), which again is greater than the expansion causing first yield of a point of the pipeline cross-section. These considerations may well obviate or reduce measures (e.g. rock dumping) to reduce pipeline expansion. However, it

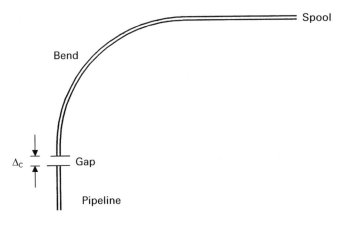

Figure 6.9   Cold-springing of pipeline bend

should be verified, first that the initial expansion can be physically accommodated without interfering with fixed structures, and second that the permanent plastic steel strain does not exceed allowable limits, which are normally put at 2.0%, provided that the girth welds are subjected to an engineering criticality assessment (ECA), see Section 8.3.2. For a permanent plastic strain not exceeding 0.3%, normal requirements to workmanship, non-destructive testing, etc. are considered adequate.

6.5.3 *Pipeline buckling*

In-service buckling of submarine pipelines can occur due to the axial compressive loads caused by thermal and internal pressure actions. If the pipeline is resting on the seabed a lateral deflection mode will prevail (snaking), whereas a vertical deflection mode will take place when the pipeline is trenched and buried (upheaval buckling). Furthermore, buckling of a free spanning pipeline may result in a downward buckling mode. When buckling occurs, part of the constrained thermal expansion is released, thus reducing the compression force in the buckled section. The resulting buckling configuration (e.g. mode, wave length and amplitude) depends to a large extent upon the frictional resistance (axial as well as lateral) between the pipe and the soil. An external force, for example, trawl board impact or anchor hooking, may initiate the buckling event.

As mentioned above, snaking will occur when the pipeline is resting on the seabed, such that the lateral restraint is less than the vertical restraint. Experimental (model) investigations of the phenomenon (Miles and Calladine 1999) have shown that once a buckle is initiated it will grow as the adjacent restrained pipeline feeds into it. Eventually, however, the first buckle lobe reaches a limit length and amplitude, and further buckling will take place in adjacent lobes, until they, in turn, stop growing, thus snaking is not likely to cause harmful deformation of the pipe. The limiting buckle size depends upon the pipe bending stiffness and the coefficients of pipe–soil friction. Indeed, the maximum strain was found to be determined by the simple formula:

$$\varepsilon = 0.425 \, D \, (\mu_a / EI)^{1/3} \, \mu / \mu_a$$

where

ε maximum bending strain
D steel pipe outer diameter
EI pipe bending stiffness
μ lateral friction coefficient
μ_a axial friction coefficient.

In the ongoing joint industry project HOTPIPE, conducted by DNV, the overall design philosophy is to allow snaking, but to control the buckling behaviour by intermittent rock dumping. The intermittent rock berms are designed to prevent axial movements, thus isolating the pipeline sections on each side from each other. The distance between the intermittent berms will allow all expansion in the corresponding pipeline section to be safely fed into a single buckle.

While allowing pipeline buckling on the seabed is a modern design practice, the above intermittent berm design approach is very conservative, and may involve the dumping of a large amount of rock. The feeding into a single buckle is obviously the most conservative assumption but, as explained above, it contradicts the physical behaviour of the pipeline on an undisturbed seabed. The formation of one buckle only requires restraints in the pipe's natural deflection, i.e. very high values for lateral friction, resulting in lateral movements similar to the deflections found when analysing the upheaval buckling phenomenon; see Section 6.5.4. Under realistic conditions a design principle that allows snaking and omits intermittent rock dumping therefore seems to be a design that may be adopted for future lines.

For an uneven seabed, the same principles of snaking apply. Typically the pipe will lift off vertically next to a span shoulder. When the lifted length exceeds the compressive Euler buckling length the plane buckling shifts from vertical to horizontal. The pipe post buckling equilibrium can then be found using the lateral buckle formulation.

6.5.4   *Upheaval buckling*

Figure 6.10 shows a pipeline section that was subjected to upheaval buckling in 1986 (Nielsen *et al.* 1990a).

Prior to this incident, upheaval buckling analyses were carried out using the same classical approach used in the vertical stability of railroad tracks. However,

Figure 6.10   Pipeline section that has suffered upheaval buckling

the assumption of a constant uplift resistance, as in the case of railroad track buckling, does not have physical relevance for the upheaval buckling deflections of a buried (or gravel covered) pipe, because of the non-linear pipe–soil interaction and decreasing soil cover during upheaval. For illustration, a comparison between the classical linear buckling model and the non-linear buckling model is presented in Figure 6.11, showing pre- and post-buckling equilibrium curves for a 12″ pipe with different imperfection amplitudes. It is seen that the classical linear analysis is non-conservative with respect to allowable fluid temperature in the pipe. In particular, the amplitude of the pipe imperfection has a significant influence on the uplift behaviour of the pipe, thus reducing the allowable fluid temperature considerably compared with the classical approach.

Furthermore, work on the pre- and post-buckling behaviours of buried/gravel covered pipelines has shown that upheaval buckling does not necessarily occur as a 'snap buckling', but that the pipe can gradually creep up through the soil/gravel due to operational variations in temperature and pressure. Thus, upheaval buckling might occur years after the pipeline has been taken into service.

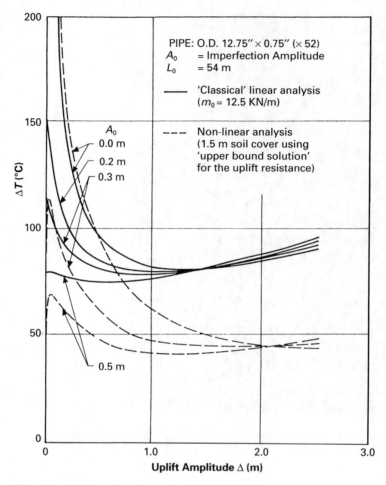

Figure 6.11 Comparison between the classical linear and a non-linear upheaval buckling analyses

Critical design parameters

A full description of the upheaval behaviour of buried pipelines is complex and requires a detailed study of all the involved parameters. In particular, the following points should be carefully reviewed when carrying out an upheaval buckling assessment:

- pipe cross-sectional properties;
- non-linear behaviour of pipe material (including temperature dependence);
- pipeline out-of-straightness;
- geometric pipe imperfections;
- soil characteristics along the pipeline route;
- non-linear pipe–soil interaction;
- varying soil cover along the pipeline;

- hydrostatic test conditions;
- operational conditions (e.g. temperature and pressure cycling).

Imperfections

Typical imperfection configurations that could be generated during the pipeline installation should be established at the design stage to assess the feasibility of employing a particular installation technique, and to identify critical imperfections in the pipeline. A conservative and realistic design imperfection is the shape of a propped pipeline due to self-weight. Thereby, the relationship between the imperfection height and the corresponding imperfection wavelength is uniquely defined by:

$$L_0 = (1152 \, EI \, A_0 / W_s)^{0.25}$$

where

L_0 imperfection wavelength
A_0 imperfection amplitude
W_s submerged pipe weight per unit length
EI pipe bending stiffness.

The submerged pipe weight should account for any additional weight on the pipe, for example, installed anodes and the risk of soil load on top of the pipe during trenching.

It is important to distinguish between foundation imperfections and geometric pipe imperfections. Foundation imperfections are characterised by variations in the vertical support conditions, and the resulting pipe configuration is governed by the bending stiffness and self-weight of the pipe. Consequently, it induces restoring bending stresses in the pipe, which will tend to straighten the pipe again should the foundation imperfection be removed. Geometric imperfections are defined by a permanent vertical out-of-straightness of the pipe centreline. In this case, no restoring bending stresses are present to 'straighten' the pipe, i.e. geometric pipe imperfections result in larger uplift displacements as compared to a pipeline with a similar foundation imperfection. Geometric pipe imperfections may be introduced in pipelines due to such problems as pipe joint misalignment, welding repairs or material yielding during the installation (particularly in the case of reeling) and trenching operations.

Pipe–soil interaction

A typical uplift resistance curve of a buried pipeline is shown in Figure 6.12. The force–displacement relationship is characterised by three discrete points. Point 1 corresponds to the submerged weight of the pipe and the weight of the soil column directly above the pipe. The displacement at Point 1 can for all practical purposes be assumed to be zero. From Point 1 to 2, the gradual upward movement of the 'imperfect' pipe section mobilises an uplift resistance due to the activation of shear stresses in the soil. It has been shown (Trautmann *et al.* 1985) that the displacement required to reach the peak uplift resistance, Point 2, can be estimated by:

$$d_2 = (0.02 + 0.008H/D)D < 0.1D$$

where

d_2 peak uplift displacement
D pipe outer diameter
H depth of burial measured to the pipe centreline.

If the upward movement of the pipe exceeds the displacement corresponding to peak uplift resistance, general shear failure occurs in the soil, resulting in decreasing uplift resistance until Point 3, where the pipe centreline has reached the surface of the seabed.

Design criteria

The following three design criteria should be taken into account when carrying out upheaval buckling analyses:

(1)   maximum allowable uplift displacement;
(2)   an adequate safety margin against snap-through buckling;
(3)   a minimum acceptable bend radius at service.

The maximum pipe uplift during temperature and pressure cycling should be limited to the 'linear' elastic deformation characteristics of the soil, i.e. between Points 1 and 2 according to Figure 6.12. Otherwise, an imperfect pipe section may be subject to an upward ratcheting effect (upheaval creep) until the overburden is insufficient to prevent upheaval buckling. Also, the build-up of residual axial compression to the side of the imperfect section during temperature and pressure cycling may further contribute to this ratcheting behaviour. As a guideline, the uplift displacement may be limited to $0.75d_2$, where d_2 is the peak uplift displacement.

In addition to the above upheaval creep design criterion, a 'snap-through' buckling design criterion should be applied to make sure that a small disturbance of the pipe does not result in a snap buckling failure. The 'snap-through' buckling criterion may be satisfied by ensuring that the width between the pre- and post-buckling equilibrium curves is not 'too small' at the specified design conditions. Typically, a width of not less than 0.1 m may be used, based upon equilibrium curves plotted in a displacement versus temperature (or pressure) plane.

To reduce the risk of localised pipe bending after an upheaval buckling failure, the minimum acceptable bend radius should not be violated. This is particularly

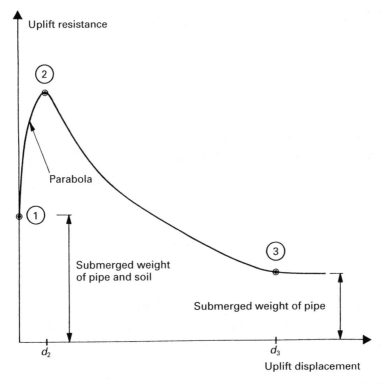

Figure 6.12 Typical uplift resistance versus pipe displacement for a buried pipe

relevant for flexible pipes, as the resulting buckling wavelength is short because of their small bending stiffness, compared with steel pipes. Thus, even though a flexible pipe can tolerate a tight bend radius, the severe pipe curvature resulting from upheaval buckling may damage the pipe, possibly resulting in a leakage.

Verification

A post-installation inspection of submarine pipelines that are susceptible to upheaval buckling should be carried out to verify that the installation specifications have been complied with. In particular, the requirements to out-of-straightness and overburden (e.g. burial depth and gravel cover height) should be verified with confidence prior to taking the pipeline into operation.

It is almost impossible to distinguish between foundation and geometric pipe imperfections when the pipeline has been trenched. However, this information may be established by comparing the out-of-straightness configurations with the as-laid and as-trenched pipeline surveys (Nielsen *et al.* 1991). Furthermore, this survey philosophy may be used to determine the depth-of-burial along the pipeline accurately. Alternatively, special intelligent pigs may be used to establish the pipe out-of-straightness.

Alternative options and remedial actions

The most common way to mitigate upheaval buckling is to provide the pipeline with sufficient overburden, for example by burial, gravel dumping, or a combination of the two (e.g. using intermittent gravel dumping). However, there are other options that can be considered to reduce the risk of upheaval buckling, including:

- an increase in lay tension;
- installation with internal pressure (flexible pipes);
- installation at elevated temperature (e.g. hot water flushing);
- an increase of thickness/density of concrete coating;
- reduction of wall thickness using strain-based design criterion;
- the use of pipe-in-pipe solutions;
- the use of pipe bundles;
- mechanical pre-tensioning of the pipeline;
- snaking the pipeline to alleviate axial loads;
- the use of mid-line expansion offsets.

If an upheaval buckling failure occurs, immediate safety measures should be taken to minimise the risk of pollution. In particular, the buckled pipe section may have been subjected to severe strain variations as a result of operational temperature and pressure cycling. Consequently, it could be important to maintain steady production to avoid fatigue-induced crack growth (Nielsen *et al.* 1990b). The repair strategy depends upon the severity of the upheaval buckling, for example, accumulated strain, resulting ovality of the pipe cross-section and exposure to third party damage. Often, the repair strategy may simply be to support the upheaved pipe section followed by mechanical protection, e.g. by gravel dumping.

6.6 Corrosion prevention and insulation

6.6.1 *General*

Corrosion is defined as a destructive attack on metal by a chemical or electrochemical reaction with its environment. The driving force is the tendency for the refined metal to return to a natural state characterised by a lower level of internal energy. In the case of linepipe steel, the iron will tend to revert to its natural state as ferrous oxides (iron ore).

Internal corrosion of pipelines depends upon the aggressiveness of the transported medium, and may be prevented by inhibitor injection, internal coating or use of corrosion resistant alloys. The lifetime of the pipeline can also be extended by introducing a corrosion allowance, i.e. an additional wall thickness over and above that needed for pressure containment. However, the extra steel is of little use in the case of pitting corrosion, which typically occurs when the corrosive medium is stagnant, or erosion, which typically occurs at high (local) fluid velocities or where scouring action may occur.

External corrosion of a pipeline in seawater is an electrochemical process. A galvanic element is created where an electric current flows between an anodic area and a cathodic area, with the seawater acting as an electrolyte. Corrosion at the anodic area will typically be localised (pitting), and a corrosion allowance offers little protection against external corrosion. Coating the steel surface protects against corrosion by creating a physical barrier between the pipe and the electrolyte, preventing oxygen from reaching the steel. Cathodic protection, on the other hand, renders the steel immune to corrosion by lowering the electrical potential, as described in Section 6.7.

Traditionally, a barrier coating is seen as the primary defence against corrosion, cathodic protection being a back-up measure against coating damage or breakdown. However, cathodic protection might also be considered the principal corrosion prevention method, the coating being introduced to reduce the necessary current consumption; see Section 6.7.3.

To ensure the proper flow of the medium through the pipeline it may be required to provide the pipe with thermal insulation, which can conveniently be combined with the external anti-corrosion coating. Alternatively, active heating of the pipeline may be an option. In such cases, however, it is important to consider the effect of temperature on the electrochemical processes involved in corrosion and cathodic protection. In theory, the reaction rate doubles for every 10°C increase in temperature, but in practice, due to scale build-up, etc., the rate 'only' doubles for every 30°C increase.

6.6.2 *Corrosion of steel in seawater*

When iron corrodes in contact with an electrolyte the neutral atoms pass into the solution through the formation of positively charged ions, leaving behind negative electrons:

$$Fe \rightarrow Fe^{++} + 2e^-$$

This reaction is termed the anodic reaction. The iron ions may further react with negative ions in the environment to produce insoluble corrosion products, typically ferrous oxides or hydroxides. The surplus electrons travel to the steel surface (through the steel), where they take part in a cathodic reaction with water and oxygen, reducing the latter to hydroxide ions:

$$2H_2O + O_2 + 4e^- \rightarrow 4\ (OH)^-$$

The dissolution of iron may continue only if the surplus electrons left behind by the anodic reaction are consumed at the same pace by the cathodic reaction. The iron dissolution rate and the concurrent consumption of electrons can be measured as a flow of current from the steel to the electrolyte. The magnitude of the current is proportional to the corrosion rate which, for unprotected steel in quiescent seawater, may be up to 0.1 mm/year on average. Locally, however, the corrosion rate on the seabed may well be significantly higher, and reaches 0.5 mm/year in the most corrosive splash zone.

As seen in the above cathodic reaction, there may be an increase in alkalinity $((OH)^-)$ at the steel surface, which may cause deterioration (saponification) of paints and other coatings.

In an acidic environment the prevailing cathodic reaction may change to hydrogen reduction due to the increased availability of hydrogen ions:

$$H^+ + e^- \rightarrow H$$

Normally the hydrogen atoms will combine to harmless molecular hydrogen (H_2), but in the presence of so-called promotors (sulphides, thiocynates, arsenic or selenium compounds) the hydrogen atoms may diffuse into the steel, where they give rise to hydrogen embrittlement. The diffusion process is enhanced if the steel is also subject to tensile stresses; see Section 3.3.3.

Hydrogen embrittlement due to corrosion can be just as dangerous to the integrity of the pipeline as the direct metal loss, especially for high strength steel and high hardness welds.

In practice the rate of corrosion is determined by the sustainability of the cathodic reaction, i.e. the availability of water and oxygen is normally the limiting factor. Hence the corrosion rate can be high near the sea level due to high oxygen availability. It may also be high at the seabed if scouring contributes by removing the corrosion products. In anaerobic soils sulphate-reducing bacteria may assist by augmenting the possible cathodic reactions. This will seldom result in general corrosion, but rather leads to localised pitting.

6.6.3 *Internal corrosion prevention*

Hydrocarbons as such are not corrosive to linepipe steel, and the prime cause of internal corrosion in pipelines is water, which may be the transported medium, or be present in crude oil or natural gas. Offshore pipelines for water transportation may be needed in connection with water injection into remote fields. For corrosion prevention such water should be de-aerated and, furthermore, biocides will have to be added, primarily to prevent pollution of the hydrocarbon-carrying formation by harmful bacteria.

Seawater will also temporarily reside in the pipeline in connection with tie-in operations and hydrostatic testing. As discussed above, carbon steel will corrode when exposed to seawater with oxygen or anaerobic, sulphate reducing bacteria (SRB). The oxygen (amounting to a maximum of 10 ppm dissolved oxygen in the water) is rapidly consumed at bare steel surfaces, and transformed into rust. In a situation with decreasing oxygen availability, the pitting tendency is very low, and the amount of corrosion will be quite modest. If the fill period is less than two to three months, the pitting tendency will remain low, the SRB not having time to develop to a degree where microbial corrosion is noticeable, and the corrosion due to oxygen reduction will be in the form of general corrosion to a depth of less than 0.025 mm. Some corrosion resistant alloys, for example, 13% Cr SMSS, are very susceptible to chloride induced pitting and crevice corrosion, so prolonged contact with seawater should be avoided.

For pipelines in operation the transported oil may contain water, either as an emulsion or as a discrete phase, and the potential for corrosion is affected by impurities such as brine, carbon dioxide (CO_2) or hydrogen sulphide (H_2S), particularly if the water is allowed to accumulate in low areas or dead spaces. In gas any water will be present as a vapour phase, but corrosion may occur if the water is allowed to condense and accumulate, particularly in the presence of CO_2 (sweet corrosion) or H_2S (sour corrosion). Thus regular pigging during operation is an effective anti-corrosion measure.

The most commonly used model of uniform CO_2 corrosion (de Waard and Milliams 1975) is normally presented as a nomogram, giving the predicted corrosion rate (in mm/yr) as a function of the partial CO_2 pressure (in bar) and the operating temperature (in °C). The original de Waard and Milliams' equation may be modified by factors taking account of the H_2S concentration, flow velocity, propensity for scaling, etc. Since the attack takes the form of general corrosion, additional lifetime of the pipeline may be bought by specifying a corrosion allowance. In some cases installation considerations or pipe size availability may lead to the adoption of a wall thickness that is greater than that needed for pressure containment during operation, in which case the excess may be counted as a corrosion allowance.

The first step of internal corrosion prevention in hydrocarbon pipelines is to avoid water, either by removing it before transportation or by keeping a flow regime that prevents water dropout. The latter option may require thermal insulation, see Section 6.6.5. Particularly susceptible to corrosion are flowlines transporting untreated well fluids, where the flow will normally be multiphase (oil, gas and/or water), and the expected corrosion rate can exceed 10 mm/yr. The problem is exacerbated by mineral impurities (sand). In addition to the abrasion of the linepipe, the sand particles increase corrosion by removing corrosion products or inhibitor films from the steel surface.

Once the operating conditions are fixed, internal corrosion is most effectively prevented by the addition of corrosion inhibitors, and the inhibitor injection programme must be tailored to the composition and temperature of the transported medium. Corrosion inhibitors are under constant development, and are reported able to reduce corrosion to a few percent. The effectiveness of the inhibitor programme should be validated by service simulating testing, and closely monitored during service (see Chapter 11). Most inhibitors may show acceptable performance on clean, polished steel surfaces, but may fail to do so on actual surfaces, such as those covered with rust or scales, etc.

As an alternative or supplementary measure, the internal surface may be provided with corrosion protection, such as an internal coating or a thermoplastic liner. The most common internal anti-corrosion coating is fusion bonded epoxy. See Section 7.3.3. However, since the FBE is applied to the pipe joints before they are assembled into pipe strings, special measures must be taken to protect the internal surface of the girth welds (see Section 8.3.3) that would otherwise be particularly susceptible to corrosion attack. This problem is avoided by spray-applying the internal coating after the welding up of the pipestring. Rapid setting, polyisocyanate cured materials are widely used for in situ coating of water mains with a minimum dry film thickness of 1 mm. The same technology could be used for pipe strings intended for offshore applications.

Alternatively, a liner (typically polyethylene or polyamide) may be inserted into the completed pipeline. Such liners are mainly used for rehabilitation of onshore pipelines, but they can also be inserted into pipe strings to be installed offshore; see Section 8.3.1. However, it should be noted that intelligent pigs, used for monitoring corrosion, may not work through a thermoplastic liner. Furthermore, gases diffuse readily through the liner material, to be stopped by the steel pipe. Upon de-pressurising of the pipeline these trapped gasses may collapse the liner. This does not represents a problem for water injection lines, but for hydrocarbon flowlines the gases will have to be vented, which can be facilitated by providing the liner with longitudinal grooves next to the steel pipe. Another possibility is to perforate the liner, a single pinhole every few metres being sufficient to prevent the build-up of pressure behind the liner. Deliberately punching holes in the protective liner may seem counter-intuitive, but tests have documented that it will not lead to significant corrosion, presumably due to the aspect ratio of the pinholes (the diameters being much smaller than the length through the liner material).

Cementitious lining materials, for example, cement paste or mortar, with or without fibres, have a long track record for onshore sewerage lines or water mains, and by extension also for marine crossings and outlets. The corrosion protection properties are due to the alkalinity of the hydrated cement, which passivates the steel. The passivation is, however, destroyed by various chemical agents, notably chlorides, and cement lining is therefore most suited for potable water lines. An interesting, recent development is to combine the cement paste against the pipe wall with a polyolefin liner towards the pipeline medium. However, as yet no hydrocarbon, let alone offshore, applications are documented.

As a last resort it may be necessary to use corrosion-resistant linepipe material (e.g. martensitic chromium steel, duplex stainless steel, austenitic stainless steel or nickel alloys), either throughout or internally only, so-called clad pipe; see Section 7.2.7.

6.6.4 *External barrier coating*

To constitute a physical barrier between the steel surface and the environment the pipe coating should be impermeable to all substances necessary for the corrosion process, particularly water and oxygen, and in practice coatings are assessed by their permeability to water. Appropriate pipeline coatings are described in Section 7.4. For the splash zone of risers, which are particularly exposed, external metal sheeting may be used.

The transport of water through a solid coating is a diffusion process that, once a steady state has been reached, is governed by Fick's first law:

$$P = -D\frac{dC}{dx}$$

where

P	permeability of water
D	diffusion coefficient
C	water concentration
x	distance.

Experience shows that satisfactory performance is obtained with a water permeability of below 1 g/m^2 per day (at ambient temperature). This requirement is satisfied by standard hot applied enamel coatings, and requires no more than a 0.1 mm coating thickness of polyolefin or polychloroprene.

Instead of using the water concentration, the vapour pressure may be used, and older textbooks cite a water permeability constant measured in g cm/cm^2/h/mmHg, i.e. the flow through a unit thickness per unit area perpendicular to the flow per unit mercury pressure. For bitumen this constant is of the order of 10–20 nanograms.

The transient flow of water through the coating is governed by Fick's second law of diffusion, and is discussed in the context of thermal insulation, see Section 6.6.5.

A particular concern when a barrier coating is used in conjunction with cathodic protection is the so-called cathodic disbonding; it is due to the hydroxides produced by the cathodic reaction, which can attack the coating and cause it to become disbonded from the steel surface. This is particularly disturbing for thick film coatings, where the disbonded coating would isolate the exposed steel from the protective current, see Section 6.7.1. Chemical pretreatment and/or epoxy primers (liquid or fusion bonded) greatly improves the resistance to cathodic disbonding, which may be documented by testing, see Section 7.4.

6.6.5 *Thermal insulation*

The proper flow of the transported medium in a marine pipeline may well require a certain minimum service temperature, and in this case the pipeline must be designed with sufficient thermal insulation and/or an active heating system. The temperature required to prevent wax or paraffin deposits, hydrate formation or water dropout normally depends upon the pressure, so thermal insulation design is intimately connected with the flow analysis; see Section 2.3. The critical temperature may be influenced by the injection of additives, such as methanol, during operation.

Thermal insulation is provided by the pipeline coating, and by the seabed soil if the pipeline is buried. The insulation properties of a material are described by the thermal conductivity, normally designated by the symbol λ or k. The thermal conductivity is measured in W/m/°C, or more correctly W m/m²/°C, i.e. the thermal energy flux through a unit thickness of the material, per unit area perpendicular to the heat flow, per unit temperature difference between the hot and the cold sides. The temperature unit °C (Centigrade) may be replaced by K (degrees Kelvin) to designate that we are talking about absolute temperatures, but this is immaterial as it is the temperature difference that matters.

The efficiency of a given insulation system is described by the heat transfer coefficient or U-value, measured in W/m²/°C, i.e. the heat loss per unit area of the insulated body. For pipelines, however, it is not obvious at which diameter the area should be measured, so it is convenient to use a total heat transfer coefficient U_p measured in W/m/°C, i.e. the heat loss per unit length of insulated pipeline. The coincidence of units makes it possible to confuse the coating heat transfer coefficient (U_p value) with the material thermal conductivity (k value), particularly since the numerical values may well be of the same order of magnitude. It does not help that some countries use the symbol k to designate heat transfer coefficients, whereas the thermal conductivity is termed λ.

For a multilayer thermal insulation it is convenient to consider the inverse of the heat transfer coefficient (sometimes called the insulance), because insulances are additive. Thus the total heat transfer coefficient U_p of a pipeline comprising a number of layers is found from the formula:

$$\frac{2\pi}{U_p} = \Sigma \ln\left(\frac{r_{oi}}{r_{ii}}\right)\bigg/k_i$$

where

$\quad U_p \qquad$ total heat transfer coefficient (in W/m/°C)
$\quad r_{oi} \qquad$ outer radius of layer Number i
$\quad r_{ii} \qquad$ inner radius of layer Number i
$\quad k_i \qquad$ thermal conductivity of layer Number i (in W/m/°C).

To this should be added the inverse of the heat transfer by radiation and convection at the internal steel surface and at the external surface of the coating, but for coated pipelines these contributions are small, and are normally neglected. The summation is taken over all the layers, including the linepipe steel, but with a thermal conductivity of 45 W/m/°C the contribution of this to the insulance is negligible. Concrete coating, on the other hand, has a thermal conductivity of 1.5 W/m/°C. If the pipeline is buried, a further contribution to the insulation is obtained from the seabed soil, which may be assumed to have a thermal conductivity of the order of 1–2 W/m/°C. The soil cover is not rotationally symmetric, but it may be included in the above formula assuming an equivalent 'outer radius' of the soil layer of:

$$r_G = H + \sqrt{H^2 - r^2}$$

where

r_G equivalent soil layer radius
r outer radius of the coated pipe
H distance from the seabed to the centre of the pipeline.

The formula presupposes a certain burial depth (at least 0.6 m). For shallowly or partially buried pipelines more accurate calculations are needed, or the insulation provided by the soil should be prudently neglected.

The thermal conductivity of hot applied enamels is typically 0.16 W/m/°C, and it is in the range from 0.2 to 0.4 W/m/°C for polyolefins (polyethylene, polypropylene and polyurethane), polychloroprene (Neoprene) and epoxy. Superior isolation properties are obtained with polyolefin foam, thus a typical polyurethane foam will have a thermal conductivity of 0.04 W/m/°C. Alternatively, hollow microspheres may by mixed into the polyolefin, and these so-called syntactic foams have thermal conductivities in the order of 0.1–0.2 W/m/°C. For more specific thermal conductivity values see Section 3.6.2.

For pipelines that are uncoated, or provided with only a thin film epoxy coating, the contributions from internal and external convectivity become significant, reducing the heat transfer coefficient, U, to approximately 100 W/m²/°C, irrespective of the steel wall thickness.

Hygrophobic syntactic foams are used without an external water barrier, and account should be taken of the dependence of the thermal conductivity on the water content. Thus a 10% water uptake typically increases the thermal conductivity by 12%. The degree of saturation depends on the water pressure and immersion time, and should be documented by tests. Apparently PU-based syntactic foams are somewhat more impermeable than PP-based ones.

The ingress of water into the coating is a diffusion process, governed by Fick's second law:

$$D\frac{\partial^2 C}{\partial x^2} = \frac{\partial C}{\partial T}$$

where

D	diffusion coefficient
C	water concentration
x	distance from the surface
T	time.

The diffusion coefficient, which is of the order of 0.01 mm^2/day, does not appear to depend on the pressure level. Calculations then show that with a 10% water content at the surface, the outer 20 mm would experience an average water uptake of approximately 5% over a period of 20 years, the remainder of the coating being practically dry (Luo *et al.* 2001).

The thermal insulation may be replaced or supplemented by active heating of the pipeline, for example, by means of electrical resistance wires embedded in the coating or in the annulus medium of a bundle or a pipe-in-pipe system. Alternatively, the pipe wall may be heated by an electric current passing through the linepipe steel and returning by a cable, piggy-backed on to the pipeline or installed separately. Another possibility is to pass the return current through the seawater by means of buried anodes or similar. This direct electrical heating (DEH) is most effective if the steel is isolated from the seawater, and if the pipeline is provided with anodes the corresponding current loss must be taken into account.

The thermal insulation system is designed to provide adequate insulation for the operational flow conditions. In the case of a shut down, the pipeline content will eventually cool down below the critical temperature unless active heating is provided. An interesting possibility for increasing the allowable shut-down period would be to insulate the pipeline with a phase change material that gives off latent heat as the cooling front moves inward.

6.7   Cathodic protection

6.7.1   *General*

Corrosion of steel in seawater is an electrochemical process, where the anodic reaction involves the loss of iron ions and the formation of ferrous oxides and hydroxides, i.e. rust. The principle of cathodic protection is to impose a galvanic element where the pipeline is the cathode.

The element may be formed actively by impressed current. Direct current is delivered from a rectifier inserted between the pipeline and permanent anodes placed on the seabed or buried in the seabed soil. Alternatively, the pipeline may be connected to sacrificial anodes made from metals that have a lower natural potential (i.e. are less noble) than steel. Sacrificial anodes are normally mounted on the pipeline (bracelet anodes), but may also be installed on the seabed (anode assemblies). The latter option is particularly suitable for retrofitting, if the pipeline anodes have been consumed or become disconnected. In any case, the protective current returns to the pipeline through the seabed and the seawater, which acts as an electrolyte.

It is of course necessary that the electric current can reach the steel surface. One could speculate that this could be impeded by barrier coatings, but tests have demonstrated (Werner *et al.* 1991) that this is not the case. Yet there is some concern over the possibility of cathodic disbonding, see Section 6.6.4. In practice, however, there is no evidence of cathodic disbonding having been the cause of corrosion on marine pipelines.

The cathodic protection system should be designed to deliver the necessary protection during the design lifetime of the pipeline. The requirements concern the protective potential as well as the protective current demand. Electrical isolation between different systems may be required, but this is not a technical necessity. Account must be taken of stray current interference, particularly from power cables.

Guidelines for cathodic protection design have been issued by a number of organisations, including Det Norske Veritas. The relevant recommendation, DNV RP B401 *Cathodic protection design*, covers impressed current as well as sacrificial anode systems. Although considered to be quite conservative in the choice of design parameters, the document is widely used by the industry, and it forms the basis for the guidance given in Section 6.7.4. The NORSOK M-503 *Cathodic protection* is similar to the DNV RP B401, whereas a radically different approach is taken by an ISO standard, ISO/DIS 15589-2 *Cathodic protection for pipeline transportation systems – Part 2: Offshore pipelines*, which is currently under development.

6.7.2 *Protective potential and current requirements*

The free corrosion potential of steel in water is approximately −0.55 V, measured against a silver/silver chloride/seawater reference electrode, which is the customary standard for marine pipelines. In what follows, all potentials are assumed to be so defined, unless explicitly stated otherwise. It is assumed that the seawater is normal ocean water (e.g. North Sea); in brackish waters the Ag/AgCl electrode potential should be corrected to reflect the actual chloride concentration.

A lowering of the potential by 0.10 V will reduce the corrosion rate by 90%, but to render the steel immune to corrosion the potential should be reduced to −0.80 V. Under anaerobic conditions, where bacterial activity may assist the cathodic reactions, the protective potential should be lowered an additional 0.10 V to −0.90 V. This will typically be the case in sediments with oxygen-consuming organic material. Furthermore, the above potential criterion is suitable for steel at ambient temperature; for service temperatures exceeding 25°C the protective potential should be lowered by 0.001 V per °C.

In terms of commonly used reference electrodes the required protective potential is:

silver/silver chloride/seawater	−0.80 V
silver/silver chloride/potassium chloride, saturated	−0.75 V
calomel/potassium chloride, saturated	−0.80 V
copper/copper sulphate, saturated	−0.85 V

The Cu/CuSO$_4$ reference electrode is commonly used on landlines, in accordance with international practice adopted by many textbooks, often implicitly. This should be kept in mind when potentials are taken from the literature.

A substantial part of the cathode reaction consists of hydrogen reduction (liberation of hydrogen atoms), which increases as the potential decreases. The liberated hydrogen may be absorbed into the steel, particularly in anaerobic environments in the presence of hydrogen sulphide. This may cause hydrogen embrittlement in high strength steels or weldments; see Section 6.6.2 above. In practice, hydrogen embrittlement is avoided in normal linepipe steels up to grade X70 if the potential is kept above (i.e. is less negative than) −1.10 V. However, for pipelines containing high strength steel, duplex stainless steel or SMSS the most negative potential that can be tolerated may have to be ascertained. Thus tests have shown that SMSS is susceptible to hydrogen embrittlement at potentials lower than −0.80 V. On the other hand, SMSS will not corrode as long as the potential is not higher than −0.60 V.

The current density necessary to protect exposed steel in seawater depends on the availability of oxygen, and is highest at the start, when the steel is polarised, decreasing as a steady state is reached. For a structure in the free flowing water of the open sea, a typical average value is of the order of 0.10 A/m^2. If the structure is buried in sediment the current demand typically reduces to 0.02 A/m^2. The required current density increases with the difference in temperature between the steel and the surroundings.

The steel of a pipeline provided with external anti-corrosion coating is in principle not exposed at all. However, even the best of coatings are subject to damage and deterioration, which increases with age. Thus the required protective current for a coated pipeline is the product of the current density on bare steel and the proportion of coating deterioration. Specific values for current densities and coating breakdown ratios are given in connection with the design of sacrificial anodes; see Section 6.7.4.

6.7.3   *Hydrogen embrittlement*

Cathodic protection results in the liberation of hydrogen atoms at the protected metal surface; see Section 6.6.2. While in 'statu nascendi' the atoms may enter into the metal surface. Once inside the metal the hydrogen atoms are free, unless trapped, to diffuse due to gradients in lattice dilation (i.e. at highly stressed/strained zones), solubility (i.e. differences in microstructure), concentration (i.e. towards the outer surface) and temperature. If too much hydrogen is present at too high a stress in a susceptible microstructure the result will be hydrogen embrittlement cracking and loss of internal integrity. This is in principle the same as that in the discussion in Section 3.3.3 on sour service resistance, but instead of being a result of the cathodic reaction in a corrosion process it is now the cathodic reaction impressed to subdue corrosion that provides the nascent hydrogen. The amount of liberated hydrogen varies with the cathode current density and the potential at the cathode. It may be high in the beginning, but polarisation effects soon reduce the cathodic reactions, due to alkalisation of the metal surface. In very general terms, less than 1% of the liberated hydrogen enters into the metal during cathodic protection in aerated seawater. In anaerobic mud with a possible high activity of sulphate reducing bacteria, hydrogen sulphides may also be present, and this may significantly increase the hydrogen absorption into the metals (Christensen and Hill 1988).

In normal situations linepipe steels up to grade ISO L450 (API 5L X65) are considered to be insensitive to hydrogen embrittlement under cathodic protection. If slow straining effects are imposed on the metal, for example if the pipe surface is gouged and dented, and the dent is inflated by the internal pressure, there is a risk of a hydrogen embrittlement type of damage (Christensen and Ludwigsen 1989). The gouging of the pipe surface produces a martensitic microstructure, which is very sensitive to hydrogen embrittlement.

Martensitic stainless steels, and to some degree also austenitic-ferritic (duplex) stainless steels (especially at welds), are also sensitive to hydrogen embrittlement. This is the reason that some operators impose restrictions on the cathodic potential on these materials. The stainless steels require only a minor potential shift in negative direction from the free corrosion potential in order to suppress corrosion. This would result in such low hydrogen atom activity at the surface that hydrogen embrittlement would not be an issue. However, the use of standard aluminium or zinc anodes to protect the pipeline (or the coupling of the pipeline to the platform structures) results in a lowering of the potential to such values that hydrogen absorption into the stainless steels may become a problem. One is advised to look carefully into the hydrogen embrittlement aspect of cathodic protection of stainless steels pipelines. One solution could be to isolate the pipeline sections to break the electron conducting path from other protected structures. The isolated pipeline may then be protected by impressed current systems targeted to keep the potential at safe levels. Iron sacrificial anodes can also be used.

6.7.4 *Sacrificial anode design*

The anodes should be designed to meet the following criteria.

- The anode mass shall be sufficient to provide the required protection during the design lifetime of the pipeline.
- The anode surface area at any time during the pipeline design life shall be sufficient to deliver the required protective current output.

For practical reasons the sacrificial anodes are designed as cylindrical bracelets, which are mounted on the pipeline. The reinforcement of any concrete coating is protected against corrosion by the alkalinity of the concrete, but to prevent any unexpected current drain there must be no electrical contact between the reinforcement and the linepipe or anodes.

The necessary anode mass is determined by the mean current requirement over the design life. For a pipeline provided with anti-corrosion coating the necessary anode surface area is determined by the required current output in the final (end of life) situation, when the anodes are consumed to a minimum and the coating deterioration is maximum. The most efficient design would only just satisfy both criteria, but in practice it will often be the surface requirement that is decisive, which means that the amount of anode material will normally be excessive.

The necessary anode mass is determined by the formula:

$$T = \frac{muC}{I_{\mathrm{m}}}$$

where

T pipeline design lifetime
m mass of anodes
u anode utilisation factor
C anode capacity
I_{m} mean current demand.

The utilisation factor is the proportion of the anode that can be consumed before the anode ceases to function properly. For bracelet anodes it is normally taken at $u = 0.8$.

The mean current demand is calculated as:

$$I_{\mathrm{m}} = \Delta_{\mathrm{m}} A_{\mathrm{p}} i_{\mathrm{m}}$$

where

Δ_{m} mean coating breakdown ratio
A_{p} surface area of the coated pipe steel
i_{m} mean protective current density on exposed steel.

The final current demand I_f (in A/km) is calculated as:

$$I_f = \Delta_f A_p i_f = \frac{I_a}{a}$$

where

Δ_f	final (end of life) coating breakdown ratio
I_f	final (end of life) protective current density on exposed steel
I_a	current output per anode
a	distance between anodes.

The last equation determines the anode sizes and spacing that can deliver the required current, the individual anode current output being:

$$I_a = \frac{V}{R}$$

where

V	driving voltage
R	anodic resistance.

So in order to design the size and spacing of sacrificial anodes it is necessary to evaluate the following parameters:

- anode capacity;
- protective current density;
- coating breakdown ratio;
- driving voltage;
- anodic resistance.

Each of the above issues is discussed below. It appears that, particularly for pipeline sections operating at elevated temperatures, it is quite important whether the pipeline is buried or exposed on the seabed. However, except for sections covered by artificial backfilling or rock dumping, pipelines are left on the seabed or in a trench, and subjected to natural sinking or backfilling. In such cases one solution, discussed below, is to estimate the degree of burial or seabed penetration. Another would be to carry out calculations for both the buried and the exposed situations, and adopt the most conservative anode design.

Anode capacity
The capacity (also called the electrochemical efficiency) depends on the anode material. For zinc alloys the capacity may conservatively be taken at $C = 700$ A h/kg

and for aluminium alloys at 2000 A h/kg. Higher values may be documented, but the DNV RP B401 recommends that 750 A h/kg and 2500 A h/kg respectively, be used as an absolute maximum. To calculate a lifetime in years from a capacity in A h/kg the number of hours per year is taken as 8760.

The capacities decrease with the temperature of the anode, and zinc is susceptible to disintegration by intergranular corrosion at elevated temperatures. Hence zinc anodes should normally not be used at temperatures exceeding 50°C, although some zinc-based alloys have been developed for higher temperature applications. For aluminium anodes the DNV RP B401 recommendation for the temperature dependency of the capacity is:

$$C = 2000 - 27(\theta - 20) \leq 2000$$

where

C electrochemical capacity (in A h/kg)
θ anode temperature (in °C).

The above values are quite conservative, considering the fact that long term testing of aluminium alloy regularly document capacities exceeding 2400 A h/kg.

Similar capacities for zinc and aluminium anodes are adopted by the NORSOK standard M-503 and the ISO/DIS 15589-2.

The temperature dependency implies that for pipelines operating at elevated temperatures zinc anodes are useless, and the capacity of aluminium anodes is substantially reduced. Experience indicates, however, that for aluminium anodes exposed to cold North Sea water no reduction in capacity is observed for operating temperatures below 60°C, thus the capacity may be taken at 2000 A h/kg. For fully buried anodes, however, it is clear that the capacity must be reduced in accordance with the temperature. The temperature at the anode surface may be estimated by an axisymmetric heat transfer calculation, the soil cover being approximated by an equivalent radius (shape factor) depending on the burial depth; see Section 6.6.5.

For partially buried anodes the situation is more complicated. Accepting the above argument, the exposed surface may be assumed to have the capacity 2000 A h/kg, but for the surface in contact with the soil the capacity should be reduced according to the temperature. The problem is that it is difficult to estimate this temperature without resorting to a full FEM simulation. Such analyses have, however, been carried out for a sample 18″ pipeline, subjected to burials of 50% and 33% (proportion of anode surface in contact with the seabed). The simulations considered the worst case scenario, with the insulating gap between the anode half shells at the 3 o'clock and the 9 o'clock positions, and revealed the following.

- The temperature of the upper anode half shell in seawater remains below 30°C for operation temperatures up to 60°C.
- For 50% burial the temperature of the lower anode half shell is close to the operating temperature (as is the case for full burial).
- For 33% burial the temperature of the lower anode half shell is significantly higher than the seawater temperature, but significantly lower than the operating temperature. Thus for a pipeline temperature of 40°C, the anode temperature may safely be taken as below 30°C, whereas this is unconservative for a pipeline operating at 50°C. On the other hand, it would be overly conservative to calculate the temperature as if the anode were fully buried.

Based upon this limited evidence a reasonably, but not overly, conservative approach for partially buried anodes is therefore:

Surface exposed to seawater: Capacity 2000 A h/kg

Surface in soil contact: Capacity according to temperature, calculated as:

$$\theta_a = \theta_s + 2\beta(\theta_p - \theta_s) \leq \theta_p$$

where

θ_a	buried anode temperature
θ_s	seawater temperature
θ_p	pipeline operating temperature
β	degree of burial (proportion of surface in soil contact).

If the actual burial depth (seabed penetration) of the pipeline is not known, a reasonable estimate for a medium hard seabed would be to assume that one third of the circumference is in contact with the soil.

Protective current density

The protective current density refers to the current per unit exposed area that is needed to achieve cathodic protection, and it varies over the lifetime of the pipeline. The initial value is the highest because it is required to polarise the bare steel but, as mentioned above, this does not apply to a coated pipeline. The mean current density refers to the steady state situation, whereas the final current density takes into account the requirement to re-polarise exposed steel in case of damage.

The design current densities depend upon the climatic zone and the water depth, decreasing with increasing water temperature and depth. DNV RP B401 offers the guidelines given in Table 6.15.

The protective current density should be increased by 0.001 A/m² per °C that the temperature of the anode surface exceeds 25°C.

Table 6.15 Design mean, i_m, and final, i_f, current densities (in A/m²) on seawater exposed steel as functions of temperature, θ, and water depth, h

		Tropical θ > 20°C	Subtropical 12°C < θ < 20°C	Temperate 7°C < θ < 12°C	Arctic θ < 7°C
i_m	$h < 30$ m	0.07	0.08	0.10	0.12
	$h > 30$ m	0.06	0.07	0.08	0.10
i_f	$h < 30$ m	0.09	0.11	0.13	0.17
	$h > 30$ m	0.08	0.09	0.11	0.13

The NORSOK M-503 and the ISO/DIS 15589-2 standards specify values of the same order of magnitude.

The current densities are substantially reduced if the structure is buried in sediment, and the DNV RP B401 as well as the NORSOK M-503 and the ISO/DIS 15589-2 standards recommend that the value $i_m = i_f = 0.02$ A/m² be used for all water depths and climate zones (corrected for elevated anode temperature as above). For a pipeline section that is partly buried, the current density may be calculated as a weighted average of the densities on buried and exposed surfaces. On the other hand, the NORSOK M-503 stipulates that the value $i_m = i_f = 0.04$ A/m² applies to 'pipelines if burial is specified'. Presumably the reference is to trenched pipelines, and the more conservative current density reflects the fact that unless the pipeline is artificially backfilled it will be only partly covered by sediment.

Coating breakdown

The coating breakdown ratio increases with time, and guidelines are provided by DNV RP B401, which specifies:

$$\Delta_m = 0.05 + 0.002 \ (T - 30)$$

$$\Delta_f = 0.07 + 0.004 \ (T - 20)$$

The above applies to the standard coating types (or equivalent); see Section 7.4:

- asphalt enamel, protected by concrete weight coating;
- three-layer polyolefin coating;
- elastomer coating.

Similar breakdown ratios are specified by the NORSOK M-503 standard. The coating breakdown is partly due to water permeation, and partly due to mechanical deterioration and damage, which is why some protection is offered by the concrete coating.

For thin film coatings (fusion bonded epoxy, epoxy paint systems) the DNV RP B401 gives formulae that depend upon the water depth as well as the number of coats and/or the dry film thickness. The NORSOK M-503 gives a single set of values for thin film coatings applied in accordance with NORSOK standards. In both standards the specified final breakdown ratio after 30 years is at least 0.4.

The above coating breakdown ratios are, however, not based upon any operational experience, and have been widely criticised as being overly conservative. The upcoming ISO/DIS 15589-2 will introduce more realistic values. The standard is still in draft form, but indications are that the specified final coating breakdown ratios for thick film coatings will range between:

$$\Delta_f = 0.01 + 0.0005 \, T \quad \text{for asphalt enamel protected by concrete, and}$$

$$\Delta_f = 0.002 + 0.0001 \, T \quad \text{for fully bonded thermal insulation systems}$$

The average breakdown ratios are found by replacing T by $T/2$. Thus the values are an order of magnitude lower than the current DNV recommendation.

Driving voltage

The driving voltage is the difference between the potential of the sacrificial anode and the steel potential needed to protect the pipeline. DNV RP B401 gives the potential of zinc and aluminium anodes buried in sediment as −0.95 V, measured against a silver/silver chloride/seawater reference electrode. If the anode is exposed in seawater the potential may be taken at −1.00 V for zinc anodes and −1.05 V for aluminium anodes. As mentioned in Section 6.7.2, the protective potential (vs. Ag/AgCl/seawater) is −0.80 V, decreasing to −0.90 V under anaerobic conditions. However, it is the former value that shall be used in designing the anodes. Thus the driving voltage may be taken as the values in Table 6.16.

For service temperatures higher than 25°C the above values should be reduced by 0.001 V per °C.

The anode potentials specified by the NORSOK M-503 standard are somewhat lower in most cases, and the same values appear to be adopted by ISO/DIS 15589-2. The resulting driving voltages are significantly higher, particularly for buried aluminium anodes; see Table 6.16.

Full polarisation of a pipeline with aluminium anodes will thus lower the potential to −1.05 V, which is below the critical value for hydrogen embrittlement of SMSS. For such linepipe steels other anode materials, such as iron, should be evaluated, or special measures be taken to avoid stress concentrations that may lead to hydrogen embrittlement.

Anodic resistance

The anodic resistance for a cylindrical bracelet anode may be calculated from the McCoy formula:

Table 6.16 Design driving voltages for pipeline anodes exposed to seawater or buried in sediment

Anode material	Environment	Driving voltage (V)	
		DNV	NORSOK/ISO
Zinc	Exposed	0.20	0.23
	Buried	0.15	0.18
Aluminium	Exposed	0.25	0.25
	Buried	0.15	0.20

$$R = \frac{0.315\,\rho}{\sqrt{A_a}}$$

where

ρ specific resistance of the surrounding medium

A_a exposed anode surface area.

The resistivity of seawater depends upon the salinity and the temperature, and in marine sediment the resistivity is higher by a factor of two to five. In DNV RP B401 curves are given for seawater resistivity as a function of temperature and salinity, and for temperate regions the values $\rho = 0.30\ \Omega$ m and $\rho = 1.5\ \Omega$ m are recommended for exposed and buried anodes, respectively. The NORSOK M-503 standard specifies only $\rho = 1.3\ \Omega$ m for anodes in sediment.

To ensure adequate protection, the distance between the individual anodes should not exceed approximately 200 m. If the pipeline is to be installed by laybarge (see Section 8.4) the spacing of the anodes must be a multiple of the individual pipe joint length, which is approximately 12.2 m. To facilitate double-jointing, adopted on some barges, the spacing should also be an even number of pipe joints.

For practical reasons related to anode manufacture the minimum anode thickness is 40 mm, and the length should not exceed 1.0 m. It is worthy of note that with a constant anode surface area per km (e.g. that determined by the required anode mass) a larger current output is achieved from smaller and more closely spaced anodes by virtue of the McCoy formula. This should be borne in mind if, for practical reasons, anodes are mounted close together (twin anodes), thus eletrochemically acting as one anode.

6.7.5 *Impressed current*

Cathodic protection by impressed current requires an external power source, delivering direct current through a transformer-rectifier. Thus the natural potential difference between two metals is not a limiting factor, as the power supply can be designed with any desired driving potential.

Typical materials for permanent anodes to be placed in seawater are:

- platinised niobium or titanium;
- lead–silver alloys;
- graphite;
- magnetite;
- silicon iron.

The last three are rather brittle and lead–silver is very soft, therefore mechanical protection of these materials is required. Platinised niobium is the best, but also the most expensive, and should be used where maintenance is difficult. Although the anodes are termed permanent (as opposed to sacrificial) they will be slowly consumed, depending upon the current demand, the lifetime being of the order of 10–20 years.

For all the materials it should be observed that the current density on the surface must not exceed specific limits:

Platinised niobium	1000 A/m^2
Lead–silver Alloys	200 A/m^2
Graphite	25 A/m^2
Magnetite	100 A/m^2
Silicon iron	25 A/m^2

The high current density on platinised niobium is mostly academic because, to avoid interference problems (see Section 6.7.7), the exit potential should not exceed 24 V. Platinum does not tolerate low frequency changes in the current load, which means that the supplied direct current should be smooth, with a ripple factor of less than 1%.

The high driving potentials needed to increase efficiency require that the anode beds shall be located at some distance from the pipeline to avoid overprotection. It must be documented that the steel surfaces that are closest to the anode do not get polarised to a level below −1.10 V; see Section 6.7.2. Thus the variation in protection along the pipeline shall not exceed 0.30 V. Instead of concentrated groundbeds it is possible to place a thin, continuous anode along the pipeline. Zinc ribbon anodes placed in low resistivity backfill may be used for this purpose. Alternatively, the anode may be made from a copper conductor coated with conductive polymer, surrounded by coke.

It is possible to design the impressed current system to be potential controlled. Reference electrodes are placed at the locations where one can foresee that the highest and the lowest protection levels will occur, and the power supply is controlled by the potential differences between the reference electrodes and the protected steel. The problem is to provide a reference electrode with sufficient long term stability. An isolated zinc anode with a slight anodic load seems to offer the best performance.

A recent development is the use of pulsed DC rectifiers, typically delivering current in the range of 300 V at 5000 pulses per second. The very brief pulses mean that there is no metal loss at the exit of the current, and the protection is more evenly distributed.

The reach of protection from an anode bed or an anode assembly is in the range 10–15 km, so cathodic protection by impressed current is not feasible for long, inaccessible marine pipelines. It is a viable option, however, for minor marine crossings, where the protection of the subsea line can be integrated into the impressed current system of the adjoining landlines.

6.7.6   *Electrical isolation*

It is not uncommon to specify electrical isolation between marine pipelines and the adjoining landlines and/or offshore platforms, mainly for contractual reasons (responsibility allocation). From a technical point of view isolation is not a necessity, as the pipeline will be protected by the nearest source. In theory a defective impressed current system on a landline could lead to premature consumption of sacrificial anodes in the surf zone, where replacement or retrofitting is next to impossible. In practice, however, the current demand of a coated landline is too low to be of any concern. At platforms it is an advantage that spool pieces, valve assemblies, etc. be protected by the platform anodes (or impressed current system) as well as by the pipeline anodes. On the other hand, the isolation of an offshore installation or a landline section provided with impressed current enables the level of protection to be assessed by measurement of the off potential.

Electrical isolation is achieved by isolation couplings (flanges or monoblock joints, see Section 7.9.2). An isolation flange has electrically isolating gaskets and, to prevent current leakage in case the flange is bridged by conductive deposits, the adjoining pipe should be provided with an isolating internal coating over a length of four times the internal pipe diameter.

Current leakage is particularly problematic if the pipeline medium is electrically conductive, which in practice means that the resistivity is less than $1.00 \ \Omega$ m. In this case it is necessary to specify an isolation spool, which is an internally coated piece of pipe between two isolation couplings. The length of the spool depends on the resistivity of the medium, and a typical rule of thumb is that it is four times the pipe diameter divided by the resistivity in Ω m.

6.7.7   *Stray current interference*

When a pipeline is placed in an electrical field it will attract stray currents, a phenomenon known as interference. Where the current enters the pipeline the potential will be lowered, increasing the risk of hydrogen formation and embrittlement. At points where the current exits the pipeline the potential will be increased, which could lead to metal loss or increased consumption of sacrificial anodes. Therefore care should be taken to avoid strong electrical fields, for example those in connection with impressed current systems; see Section 6.7.5.

By far the biggest concern, however, is related to subsea DC monopolar power cables, with a constant direction of the current, which returns via the water and the seabed. The induced electrical fields are particularly strong in the vicinity of the anode and cathode beds, and the current pickup of nearby pipelines should be analysed.

6.8   Bends, components and structures

6.8.1   *General*

Pipeline components are structural parts that are connected to the pipeline, i.e. they carry the product pressure, but are forged or otherwise manufactured using different processes than those of linepipe production. Thus, in this book, bends are not considered to be components, as they are often manufactured from linepipe (motherpipe), as are spools and pup pieces. This is in contrast to DNV, for example, which has clarified that the rules for linepipe do not cover bends, which should be considered as hot formed components. It is common design practice, however, to apply the linepipe design criteria (see Section 6.2), to bends as well, if the bends are forged.

Components include fittings and special components, notably valves. Fittings comprise flanges, hubs, tees, wye pieces, etc., i.e. basically passive parts of the pipeline system. Special components are those that may be actuated in connection with the pipeline operation, such as valves that may be closed, isolation couplings that may be short-circuited, facilities that may dispatch or receive pigs, etc.

Structures in this context denote the steel skids and covers supporting and protecting valve assemblies and similar components on the seabed.

Risers and expansion offsets are treated in Section 7.9.2.

6.8.2   *Fittings*

Flanges may be full bore, used as an alternative to hyperbaric welding for offshore connection (tie-in) of the pipeline to risers, expansion offsets, valve assemblies, etc., or they may be blind flanges used for capping off future connections to spur lines, bypass loops, etc. Such branch line connections may be perpendicular to the pipeline (tees) or oblique (wyes), and the bore of the branch line may be the same as or smaller than the main pipeline. Small bore connections to the pipeline are usually performed by means of olets; see Section 7.9.4.

Branch connections (tees, wyes) constitute the nodes of the pipeline system, and they are normally forged steel components. A typical hot-tap tee is shown in Figure 8.19. The material is specified in accordance with the transported medium (e.g. sweet or sour service). The loads on the component are identified by analysis of the pipeline system, typical load cases being hydrotesting and operation. Verification may be carried out by finite element analysis, complying with a recognised standard, e.g. ASME B31.8 and ASME BPV Section VIII, Division 2, Appendix 4, which is based upon the absolute difference between the principal stresses (Tresca criterion).

Olets, hubs, flanges, bolts and gaskets are in principle designed the same way, but they will normally be off-the-shelf items, specified in accordance with pressure rating and internal environment, and provided with the corresponding certificates in compliance with recognised standards. In a normal flanged connection the bolt forces are transferred through the gasket, but in recently developed compact flanges, standardised by NORSOK, the forces are transferred directly through the flange faces, the seal being located in a groove. The bolts and seals are thus not subjected to the pipe system loads, which means that the connection can be made lighter as well as tighter. Compact flanges have been used with success on risers and subsea installations in the North Sea.

Paint systems are usually used for corrosion prevention, in combination with cathodic protection. Sacrificial anodes would not normally be attached directly to the component, but additional anodes are placed on the adjoining pipeline or structure, to which the component is electrically connected.

6.8.3 *Valves and other components*

Pipeline valves are typically of two types: non-return flapper valve types (check valves or emergency shut-down valves), which respond automatically to a pressure loss on one side of the valve, or ball valves, which require intervention to be opened or closed, either manually or by a valve actuator. Other types, such as butterfly valves, are not used in marine pipelines, because pig passage requires a full bore over the entire length.

For complete isolation, for example of a platform from an export pipeline, it is customary to install a valve assembly, consisting of a ball valve and a check valve in series, or alternatively a single, remotely operated, subsea isolation valve (SSIV), which can also be used on incoming pipelines. Particularly for gas pipelines this may be a legislative requirement, but subsea isolation valve assemblies can also be justified by risk analysis; see Section 5.6.1. Isolation valves may be incorporated into the riser base or installed as a separate structure some distance from the platform.

Flapper valves on gas pipelines are susceptible to vibrations induced by gas flow turbulence. To ensure an adequate lifetime of the flapper the forced response from the induced turbulence frequency should be analysed and compared with the natural frequency of the swinging flapper, and the durability of the flapper bearings should be documented.

For electrical insulation an isolation coupling may be installed. This may be a specially designed flange connection, provided with electrically isolating gasket and bolting, or an integral component (monoblock) to be welded into the pipeline.

All valves and isolation couplings are specified in accordance with pressure rating and internal environment (e.g. sour service).

Pig launchers/receivers are designed in accordance with the principles and safety format adopted for the pipeline. A receiver (pig trap) is shown in Figure 8.24.

6.8.4 *Structures*

Protective structures are required for subsea pig launching/receiving facilities as well as valves; see Figure 5.10. The structures should be designed to resist loads from fishing equipment (overtrawling) and/or dropped objects (close to platforms).

Any obstacle on the seabed is likely to cause extensive scouring, particularly in shallow water with high wave activity, and the hydrodynamic stability of the structure may be ensured by anchor piles and scour protection. The design of the latter may well require model testing as well as mathematical modelling. In deep water structures might simply be resting on the seabed; see Figure 11.1.

Long-term integrity is achieved by cathodic protection, usually by means of stand-off sacrificial anodes mounted on the structure. In Figure 5.10 an anode can be seen under the top horizontal member; see also Figure 11.1.

Chapter 7
Fabrication

7.1 Introduction

The principal building block of any marine pipeline is the pipe joint, which is an approximately 12.2 m (40′) section of steel tube (called linepipe). Before being assembled into pipe strings and installed on the seabed the individual pipe joints are provided with some or all of the features listed below, as required by the design:

- internal coating;
- external anti-corrosion coating;
- thermal insulation;
- sacrificial anodes;
- concrete weight coating.

A typical section in a coated pipe joint is shown in Figure 7.1, and Figure 7.2 shows a stack of coated pipe joint ready for shipment to the installation contractor.

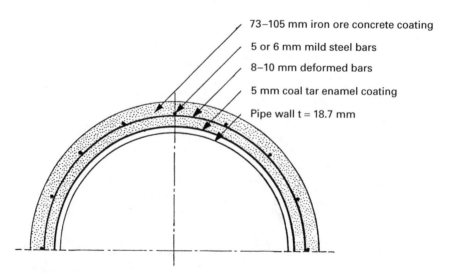

73–105 mm iron ore concrete coating

5 or 6 mm mild steel bars

8–10 mm deformed bars

5 mm coal tar enamel coating

Pipe wall t = 18.7 mm

Figure 7.1 Typical cross-section of a coated pipe

Figure 7.2   Coated pipe joints stacked for shipment

In addition to the straight pipe joints, a marine pipeline may include hot formed bends, as well as components, such as flanges, tees, wyes, valves, isolation couplings, pig launchers/receivers, etc.

This chapter describes the procedures and requirements related to the fabrication of pipe joints, bends and components.

7.2   Linepipe production

7.2.1   *General*

Steel making and subsequent pipe manufacturing may seem to be a relatively simple processes to the non-metallurgist, and it is a kind of contradiction that it takes a lot of expertise in metallurgy, chemistry and physics in order to realise how complicated it really is. It is a delicate economic and difficult technical decision to 'design' up-front the chemistry and subsequent steel making processes, refining steps, casting sequence, rolling scheme and pipe forming procedure that result in the end product meeting the specified, narrow properties for demanding applications. One mistake may result in hundreds of tonnes of steel being scrapped or re-routed to less demanding applications. Introductions to the difficult task of steel production can be found in the ASM *Handbook No 1*, published by ASM International, The Materials Information Society (formerly known as the American Society of Metals).

In the ideal world, pipeline engineers just specify the properties that they need in their design. These will typically be the pipe diameter, the wall thickness, and the required mechanical properties that allow the required system pressure to be carried in a safe manner, together with good weldability and geometrical tolerances. They need not consider the difficulties involved in changing iron ore or re-cycled scrap into modern, high quality, steel pipes. They may ignore the complicated liquid metallurgy processing steps aiming at refining the steel, controlling sulphur, phosphorus, carbon, oxygen, nitrogen and hydrogen, reducing inclusion sizes and population, preventing re-oxidation, ensuring uniform fine grain solidification, etc. They may expect the use of micro-alloying elements like niobium, vanadium, titanium and boron to result in superior toughness, strength and weldability, and not consider the intricate solid metallurgy processing steps needed to achieve the end result. Modern linepipe manufacture by Thermo-Mechanical Controlled Processing (TMCP) involves grain refinement, precipitation hardening and deformation strengthening through consecutive rolling at decreasing temperatures close to the austenite to ferrite transformation temperature, and possibly supplemented by the use of accelerated cooling.

The pipeline engineer should, however, be aware that the engineering properties in large tonnage production might vary, to say the least. This may result in catastrophically low toughness properties, substandard strength properties and impaired weldability, if well documented manufacturing procedures are not followed in all production steps, from steel making to the final pipe manufacturing, or if the implemented quality assurance plan is insufficient.

7.2.2 *Standardisation*

Simultaneously with the development in steel and pipe making practices, the pipeline industry has joined forces in developing international standards and norms to help specify the end product in unambiguous terms. The most used standards in the last half of the 20th century were the American standard API 5 L and the German Norm, DIN 17 172. These standards and a recently developed Euro Norm standard, EN 10208 were incorporated in a common international standard developed under the auspices of the International Standardisation Organisation (ISO). By the charter entrusted in ISO the common standard ISO 3183 should eventually replace existing national or local norms and standards. The ISO standard 3183-1 (requirement class A) corresponds in general to the API 5 L. The ISO 3183-2 (requirement class B) imposes additional requirements on toughness and non-destructive testing, and is in general usable for transmission pipelines. The ISO 3183-3 specifies the technical delivery conditions for steel pipes for pipelines of requirement class C, i.e. particularly demanding applications, such as sour service, offshore service and Arctic service. The ISO standards provide minimum specifications, with options for more stringent requirements regarding chemical composition, mechanical properties and test methods, as well as testing extent and pipe geometry. Recognised pipe mills should have no technical problems in meeting these requirements. It is of course possible to use even tighter specifications on specific properties to suit special purposes or imposed national requirements, but this may be at the expense of reduced competition and increased costs. Examples of tightening are:

- identical minimum requirement in longitudinal and transverse direction in terms of mechanical properties for deep water pipe projects;
- increased impact toughness requirements;
- extended use of fracture mechanics for pipes in critical applications such as in environmentally sensitive or populated areas.

Fracture mechanics principles are brought in if accumulated plastic straining of the pipe material during pipe laying or trenching operations exceeds 0.3%, or if the pipe is exposed to pipe reeling (>2%) and alternating yield.

Linepipes delivered according to ISO 3183-3, and summarised in Table 7.1, are characterised as the following products:

- seamless pipe (S);
- high-frequency welded pipes (HFW);
- submerged-arc welded longitudinal seam (SAWL);
- submerged-arc welded helical seam (SAWH).

Seamless pipes are also designated SML (e.g. by DNV OS-F101).

7.2.3 *Seamless pipes*

Seamless pipe manufacture is achieved by one of the four main processes described below. In essence all the processes consist of making a relatively short, thick-walled hollow work piece that is subsequently massaged or rolled into long pipes. The first three methods make use of the piercing roll method; see Figure 7.3, originally developed by Mannesmann, while the last method applies a piercing press; see Figure 7.4. The range of material for all processes comprises carbon steel as well as low, medium and high alloy steels, including corrosion resistant alloys.

Pierce and pilger rolling is most faithful to the original process developed by the Mannesmann brothers and is still known as the Mannesmann process. It covers typically the diameter range from 100 to 700 mm with wall thicknesses from 4 to 150 mm.

Continuous mandrel rolling uses a modern realisation of the Mannesmann piercing process. Mandrel rolling is a high capacity process producing seamless pipes from 25 to 180 mm in diameter and wall thickness from 2 to above 30 mm.

Plug rolling and the related *hot rotary expansion process* also make use of modern piercing roll equipment. Plug rolling allows pipes to be produced

Table 7.1　Type of pipes in ISO 3183-3

Type of pipe	Starting material	Pipe heat treatment	Symbol for heat treatment
Seamless (S)	Ingot or billet	Normalised during pipe forming	N
		Normalising	N
		Quenching and tempering	Q
		Normalising	N
		Quenching and tempering	Q
High-frequency welded (HFW)	Normalising rolled strip	Normalising weld area	N
	Thermomechanically rolled strip	Heat treating weld area	M
		Heat treating weld area and stress relieving (entire pipe)	M
	Hot rolled or normalising rolled strip	Normalising (entire pipe)	N
		Quenching and tempering (entire pipe)	Q
		Stretch reduced under normalising condition	N
		Cold formed and subsequently thermomechanically processed	M
Submerged-arc welded (SAW)	Normalising rolled plate or strip	–	N
• Longitudinal seam (SAWL)	Thermomechanically rolled plate or strip	–	M
• Helical seam (SAWH)	Quenched and tempered plate	–	Q
	As rolled, normalised or normalising rolled plate or strip	Quenching and tempering (entire pipe)	Q
	As rolled plate or strip Normalised or normalising rolled plate or strip	Normalised during pipe forming	N

without the need for reheating. It is used for production of pipes in the 170 to 700 mm diameter range with wall thicknesses of between 5 and 30 mm.

Pierce and draw (Eberhardt) produces pipes and hollows in the diameter range 180 to 1200 mm in thickness, and ranging from 20 to 270 mm.

7.2.4　Welded pipes

HFW Pipes (Longitudinal weld)

The high-frequency weld process (see Figure 7.5) uses strips (skelps) that are successively dished and shaped into a U form, followed by forming through closed rim rollers into an open, O shaped pipe in a continuous roll forming

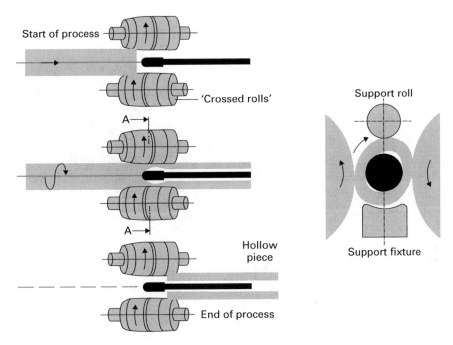

Figure 7.3   Schematic presentation of roller piercing (reproduced after (Krass *et al.* 1979)

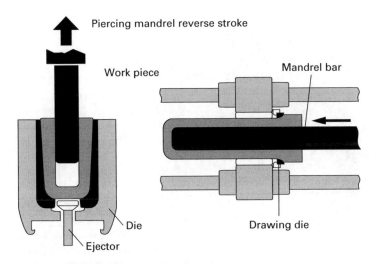

Figure 7.4   Schematic presentation of the pierce and draw process

process. The abutting side faces are welded with one longitudinal seam without the use of filler metal by applying high frequency current (by induction or conduction); see Figure 7.6. The forming may be followed by hot stretch, cold expansion or sizing to obtain the required dimensions. The HFW process is

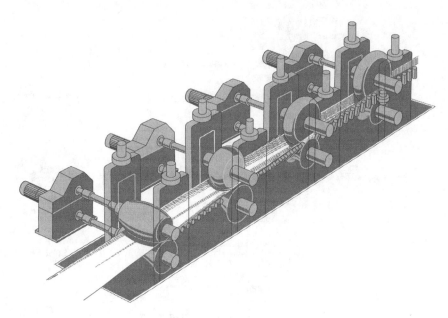

Figure 7.5   Continuous roll forming

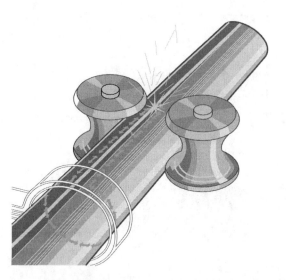

Figure 7.6   High-frequency weld (HFW) process. The electric voltage induced by the coil inductor around the pipe generates a high-frequency alternating current (~200 kHz). The current flows along the abutting edges around the contact point where it generates the required welding temperature in a narrow zone

available up to around 700 mm (28″) in diameter and wall thickness up to around 18 mm, but offshore experience with dimensions greater than OD 400 mm (16″) and WT of 16 mm is limited. As the confidence in the product increases, there may be a wider use of HFW pipes in the years to come.

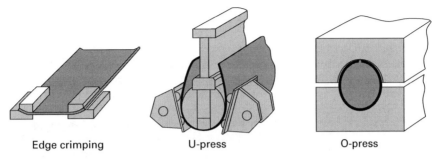

Edge crimping   U-press   O-press

Figure 7.7   Pipe forming by the UOE process

SAWL pipes (Longitudinal weld)

The pipe forming process in submerged-arc welded longitudinal seam pipes produces pipes up of to 1500 mm diameter and wall thicknesses of up to and above 30 mm. SAWL pipes can be divided into the following processes:

- UOE;
- COE;
- 3-RF;
- continuous roll-formed processes.

In the UOE process (see Figure 7.7) the plate strips are first crimped along the edges, then pressed into a U shape, followed by an O pressing. The abutting faces are welded from the inside and the outside by submerged-arc welding. The welded pipes are mechanically expanded to enhance the pipe geometry and to release the weld residual tensile stresses in the hoop direction.

In the COE process the half width of the plate strip is gradually formed to a C by press bending, starting from the edge. The same process is then repeated from the other plate edge to produce an O shape, which is subsequently welded by submerged-arc welding and mechanically expanded as above.

Three-RF (3-roller forming) is akin to a normal rolling process used in the production of shells for cylindrical vessels, but it is used here to produce long bins that are welded. The roundness obtained is very good and mechanical or hydraulic expansions are used mainly to stress relieve the weld zone.

Continuous roll-forming is equal to the HFW process (see Figure 7.5) but the welding process is submerged-arc welding. The pipes may be mechanically or hydraulically expanded.

SAWH pipes (Spiral weld)

The starting materials are strips or steel bands. Figure 7.8 illustrates how spiral weld pipes are formed by roller forming at an oblique angle to the longitudinal axis of the strip/band. By changing the angle and the curvature it is possible to

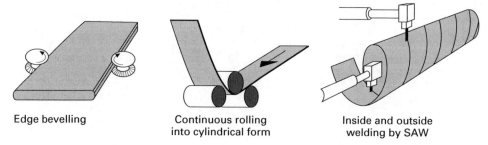

Edge bevelling   Continuous rolling   Inside and outside
                          into cylindrical form   welding by SAW

Figure 7.8   Spiral welded pipes

produce pipes of different diameters. The production range covers up to and above 2500 mm in diameter and thicknesses up to 30 mm.

7.2.5   *Specified properties of linepipe steels*

Linepipe steels are designated by L (for Linepipe material) followed by the specified minimum yield strength (SMYS) in MPa. This has been chosen to be in agreement with previous standards and to avoid misunderstanding. Table 7.2 compares the new designations with those of API 5 L, which are based on yield strength in ksi.

In ISO 3183-3, the SMYS value is followed by a suffix to classify the heat treatment condition and an additional letter S to distinguish the material as specially produced for sour service. The complete classification and designation of the steels included in ISO 3183-3 is shown in Table 7.3, and Tables 7.4 and 7.5 summarise the mechanical properties, respectively the steel chemistry (for welded pipes).

Pipeline systems shall have adequate resistance against initiation of unstable fracture. This goal is achieved by specifying minimum impact toughness values at low temperatures. Resistance against brittle fracture is obtained by ensuring that the transition temperature from brittle to ductile behaviour is below the minimum design temperature. The required minimum impact toughness values at

Table 7.2   Steel grade comparison

Steel grades in ISO 3283-3	Steel grades in API 5 L
L245 . . .	Grade B
L290 . . .	X42
L360 . . .	X52
L415 . . .	X60
L450 . . .	X65
L485 . . .	X70
L555 . . .	X80

Table 7.3 Classification and designation of steels in ISO 3183-3

Heat treatment condition	Steel class in accordance with ISO 4948-1 and 2	Steel name*
Normalised or normalising formed	Non-alloy special steel	L245NC(S) L290NC(S) L360NC(S)
Quenched and tempered	Non-alloy special steel Alloy special steel	L290QC(S) L360QC(S) L415QC(S) L450QC(S) L485QC(S) L555QC
Thermomechanically formed	Non-alloy special steel	L290MC(S) L360MC(S) L415MQC(S)
	Alloy special steel	L450MC(S) L485MC(S) L555MC

* In the steel name: L designates Linepipe; N, Q and M refer to heat treatment condition; C refers to requirements class C (in ISO 3183-1 and 2 the letter should be 'A' and 'B' respectively) and S refers to sour service materials.

Table 7.4 Required mechanical properties of steel pipes in ISO 3183–3

Steel name	Pipe body (seamless and welded pipes ≤25 mm WT)				Weld seam HFW and SAW	
	Yield strength, $R_{t0.5}$ (MPa)	Tensile strength, R_m (MPa) minimum	Ratio, $R_{t0.5}/R_m$ maximum	Elongation, A (%)	Tensile strength, R_m (MPa) minimum
L245	245–440	415	0.90	22	415
L290	290–440	415	0.90	21	415
L360	360–510	460	0.90	20	460
L415	415–565	520	0.92	18	520
L450	450–570	535	0.92	18	535
L485	485–605	570	0.92	18	570
L555	555–675	625	0.92	18	625

minimum design temperature below 10°C (according to ISO 3183-3) are shown in Table 7.6.

For high pressure gas and mixed gas and liquid pipeline systems it is desirable also to ensure against propagating fracture (long-running shear fracture). This can be achieved by any of the following measures:

Table 7.5 Chemical composition for thermomechanically processed, welded pipes according to ISO 3183-3

Steel name	Maximum content (%)									CEV§ max.	Pcm" max.
	C	**Si**	**Mn**	**P**	**S**	**V**	**Nb**	**Ti**	**Other**		
L290MC	0.12	0.40	1.35	0.020	0.010	0.04	0.04	0.04	*	0.34	0.19
L290MCS	*0.10*		*1.25*		*0.002*						
L360MC	0.12	0.45	1.65	0.020	0.010	0.05	0.05	0.04	†	0.37	0.20
L360MCS	*0.10*		*1.45*		*0.002*						
L415MC	0.12	0.45	1.65	0.020	0.010	0.08	0.06	0.06	†, ‡	0.38	0.21
L415MCS	*0.10*		*1.45*		*0.002*						
L450MC	0.12	0.45	1.65	0.020	0.010	0.010	0.06	0.06	†, ‡	0.39	0.22
L450MCS	*0.10*				*0.002*						
L485MC	0.12	0.45	1.75	0.020	0.010	0.010	0.06	0.06	†, ‡	0.41	0.23
L485MCS	*0.10*				*0.002*				*Mo ≤ 0.35*	*0.39*	*0.22*
L555MC	0.14	0.45	1.85	0.020	0.010	0.010	0.06	0.06	by agreement		

* Cu ≤ 0.35; Ni ≤ 0.30; Cr ≤ 0.30; Mo ≤ 0.10; B ≤ 0.0005
† Cu ≤ 0.50; Ni ≤ 0.50; Cr ≤ 0.50; Mo ≤ 0.50; B ≤ 0.0005
‡ The sum of V, Nb, Ti shall not exceed 0.15%.
§ Carbon equivalent $CEV = C + Mn/6 + (Cr + Mo + V)/5 + (Cu + Ni)/15$.
" Parameter crack measurement $P_{cm} = C + V/10 + Mo/15 + (Cr + Mn + Cu)/20 + Si/30 + Ni/60 + 5B$.

Table 7.6 Charpy V-notch impact toughness requirements for avoidance of brittle fracture

Grade	Average of three test specimens transverse/longitudinal (J)	Minimum individual value transverse/longitudinal (J)
L245	27/40.5	22/33
L290	30/45	24/36
L360	36/54	30/45
L415	42/63	35/52.5
L450	45/67.5	38/57
L485	50/75	40/60
L555	56/84	45/67.5

Test shall be performed:
  at project design temperature –10°C for WT ≤ 20 mm
  at project design temperature –20°C for 20 < WT ≤ 30 mm
  at project design temperature –30°C for WT > 30 mm.
Impact energy values apply to transverse full size test piece.

- periodic use of mechanical crack arrestors or sections of heavy wall pipes (to lower the stress);
- specification of low transition temperature and adequate impact toughness values (higher than those needed to ensure against brittle fracture);
- specification of drop weight tear test (DWTT) shear fracture area of minimum 85% at −10°C or minimum design temperature for the project, whichever is the lowest.

Detailed information and guidelines for the selection and testing of material for sour service, and for acceptance criteria, can be found in the European Federation of Corrosion guidelines and in ISO 15156-1 and 2.

SSC and HIC tests may be used to qualify the steel material for sour service (see Section 3.3.3 for a definition of the hydrogen-related damage phenomena associated with sour service). SSC testing is used mainly during qualification of manufacturing procedures, whereas HIC testing is used as a general quality assurance test during procurement. These tests are incorporated in ISO 3183-3. The tests consist of exposing small samples of the pipe material to a specific reference environment, typically 5% sodium chloride acidified to pH 2.8 by addition of acetic acid and saturated with 1 bar H_2S. For specific projects, the test environment may be chosen to reflect the worst-case situation in the expected service. In SSC tests the samples are stressed (usually to 80% of specified minimum yield strength), and the sample is expected to pass a 720 hour test without cracking. The HIC tests are unstressed, and the samples are evaluated after 96 hours exposure by metallographic examination for possible internal cracking and blistering.

SSC resistance is basically controlled by maintaining hardnesses below 250 HV, and by using pipes of strength grade not higher than grade L450. It is a prerequisite that the pipes are free from residual tensile stresses on the inside. Higher strength grades are considered more susceptible to SSC, and extended qualification test programmes are usually envisaged.

HIC resistance is facilitated by controlling the steel cleanliness and the size and distribution of non-metallic inclusions. A low sulphur content is essential, but also the prevention of banded microstructures (bands of pearlite) and the segregation of low temperature transformation products (bainite, martensite) is important.

SOHIC and SZC are relatively uncommon phenomena in practice, but should be considered if the pipes are highly stressed (applied and residual) during service.

It should be noted that the DNV OS-F101 adopts linepipe steel designations that are slightly different from those of the ISO 3183-3. Thus, for example, SML 450 I S designates seamless pipe with SMYS 450 MPa, subjected to NDT level I, and meeting requirements to sour service (Suffix S), whereas SAWL 450 I DFU would be longitudinal weld SAW pipe with SMYS 450 MPa, subjected to NDT level I, and meeting special requirements to dimensional tolerances (Suffix D), fracture arrest (Suffix F) and high utilisation (Suffix U). For easy reference it is also customary to use the traditional API designations, where, for example, X65 denotes linepipe with a yield stress of 65 000 psi (approximately 450 MPa).

7.2.6 *Hot formed bends*

Bends for marine pipelines are in general made in a bend machine with induction heating of a narrow zone just ahead of the bending matrices. As the pipe is pulled round by the matrices the heating zone stays at the position of maximum bending moment. External and internal cooling may be applied to obtain a proper shape and to prevent changes in material strength, etc.

The bends may be hot formed from spare pipes from the pipeline, provided that the material in the final delivery conditions will still fulfil the required dimensional and mechanical properties and maintain the required sour corrosion resistance. It is quite normal to use specially produced seamless 'motherpipes' for hot-formed bends, but also longitudinal weld motherpipe may be used.

The bend radius is normally specified as five times the linepipe diameter (5D bends) to allow passage of inspection pigs, but 3D bends are also considered piggable. Sharper bends (1.5D or even 1D) may be incorporated in non-piggable components like tees. Such items are designated elbows, and would normally be forged rather than hot formed; see Section 7.9.1.

Hot formed bends are produced with heavier walls to counteract the possible wall thinning along the neck of the bend. The amount of thinning experienced at the extrados of the bend is typically 8–9% for 5D bends and 11–13% for 3D bends.

To facilitate welding to adjoining linepipe or components, bends are normally provided with tangent lengths, and it is tempting to use this to reduce or obviate a pup piece. However, experience with geometrical variations of hot formed bend tangents indicates that it is advisable to reduce the tangent length to the minimum required for girth welding (typically 100 mm), allowing approximately 50 mm for machining and bevel preparation.

The chemistry of hot formed bends may be selected so as to either avoid post bend heat treatment or benefit from such heat treatments. It is of paramount importance that the material is easily welded to the pipe, and that the weld zone properties are comparable to other girth welded joints in the pipeline.

Guidelines for the manufacture and testing of bends are given in DNV OS-F101.

7.2.7 Cladding, lining and weld-overlaying

The concept of using carbon steel clad with a relatively thin layer (usually 2–3 mm) of corrosion resistant alloy (CRA) is well established. It has been used for vessels, separators, tanks, etc. for more than 50 years in the chemical, petrochemical and oil refining industries, and it has also gained use within the oil and gas pipeline industry over the last 20 to 30 years (Denys 1995).

According to the general terminology, cladding refers to steel pipes with a corrosion resistant layer forming a metallurgical bond with the steel. Roll bonding, explosive bonding followed by rolling, hot isostatic pressure assisted sintering/diffusion bonding and centrifugal casting may obtain the metallurgical bond. The CRAs need to be solution annealed after the hot forming operations in order to optimise their corrosion resistance. By choosing CRAs that can be solution annealed at temperatures below 1000°C it is possible to retain sufficient strength in the backing steel and to avoid grain growth and loss of ductility.

Linings, on the other hand, refer to mechanical bonding obtained by expansion of a liner on to the inside of the pipe by hydraulic or mechanical expansion or by explosive forming. These processes do not require heat treatment after lining because the lining can be used in solution annealed condition. It is therefore possible to use TMCP materials as backing steel, thus benefiting from increased strength and toughness and reduced wall thickness.

The most widely used cladding and lining alloys in the oil and gas pipeline industry are stainless steel AISI 316 and nickel alloy 825, but alloy 625 and C276 (Hasteloy) are also used.

Weld-overlays are mainly used for flange faces, valve bodies, tees and wyes and for shorter flowlines. The dominant weld-overlay materials are alloy 625 and C276.

Table 7.7 (Denys 1995) gives an overview of the available plate and pipe sizes according to various manufacturing methods. The plates can be used to produce longitudinally or spirally welded pipes.

Special procedures have been developed to allow field jointing by welding. To avoid using different welding consumables and techniques for the two layered pipes it is sometimes seen that abutting bevels are buttered by overlay welding, or that solid CRA sleeves have been shop welded to the pipes. The latter leaves some degree of freedom in adjustment and bevelling in the field.

Various corrosion tests will be requested for product qualification. They may vary from project to project, but include one or more of the tests prescribed in ISO 15156-2 for corrosion resistant alloys, i.e. testing for resistance to sulphide stress cracking, pitting corrosion (ASTM G48 in 10% $FeCl_3$), intergranular corrosion (ASTM A262, B, C or E), and stress corrosion cracking.

7.2.8 *Testing during pipe manufacture*

The pipe mill is asked to present a detailed manufacturing procedure specification, and describe a qualification test to demonstrate that the pipe mill is able to produce the pipes to the requested quality. Recognised and experienced pipe mills may avoid qualification testing by documented reference to previous production of similar pipe orders. The qualification testing may involve the following:

- mechanical requirements for both base material and weld zones, i.e.
  - □ yield and tensile strength, R_e and R_m in MPa
  - □ elongation, $A_5\%$
  - □ impact toughness values; Charpy V and Batelle drop weight tear test
  - □ fracture toughness values, (if specified)
  - □ bend test (SAW)
  - □ flattening test (HFW)
- macro- and metallographic examination
- chemical composition of base metal and weld metal, including all elements deliberately added

Table 7.7   Available plate and pipe size according to the various cladding and lining manufacturing methods

Product	Wall thickness (mm)	Width/ Diameter (mm)	Max. length (m)
Roll bonded plate	6–200 mm Cladding 1.5 mm – up to 40% of total wall thickness	1000–4450	16.5
Explosive bonded plate	Cladding 1.5–25 mm Base ≥3 times the cladding	50–3500	5
with hot rolling	No limit to base thickness	1000–4450	14
Overlay welded plate	Clad layer >2.5 mm Base metal >5 mm	Limits by access of equipment	Limits by access of equipment
Longitudinally welded clad pipe	Total wall 6–60 mm Clad layer >1.6 mm	100–1626	6–18
Lined pipe (thermal shrink fit)	Total wall 5–30 mm Liner 2–5 mm	100–610	9.6–12 depending on size
Lined pipe (explosive mechanical bond)	Base pipe wall >5 mm Liner 2–5 mm (depending on diameter)	50–400	12
Seamless clad pipe (extruded and processed in plug/mandrel mill)	Total wall 6–30 mm Liner >2 mm	50–600	9–12 depending on size
Seamless clad pipe (metallurgical bond by explosive)	Base pipe 2–20 mm Liner >1.6 m	50–250	3–5
Seamless clad pipe (cold rolling)	4–25 mm	20–240	2–16
Centri-cast pipe	Total wall 10–90 mm Clad layer >3 mm	100–400	4–6
HIP clad pipe or fittings	Total wall >5 mm Clad layer >2 mm	25–1000	2
Weld overlay fittings	Base material >5 mm Clad layer >2.5 mm	25 minimum	Depending on size

Note: Not all combinations of wall thickness/diameter/length may be possible.

- non-destructive testing of pipe body, weld zone and pipe ends
- simulated field weldability testing
- residual magnetism in pipe ends
- pipe dimensions, i.e.
  - ☐ length
  - ☐ diameter
  - ☐ wall thickness

    ☐ ovality

    ☐ straightness

    ☐ weight

- corrosion testing, i.e.

    ☐ HIC tests

    ☐ SSC tests

    ☐ SOHIC test

- other relevant testing according to pipe material, e.g. CRAs or clad pipes

    ☐ Pitting corrosion

    ☐ Intergranular corrosion

    ☐ Stress corrosion cracking.

It may be agreed to perform the manufacturing procedure testing as a part of the initial production at the manufacturer's risk.

During production the same testing as mentioned above (except field weldability) is performed at a prescribed frequency. Some tests, such as visual inspection, hydrostatic testing, wall thickness at pipe ends, weight and non-destructive tests of pipe body, weld and pipe ends, are performed on every pipe. Mechanical properties and chemistry will be checked once per 50–100 pipes or per heat (same charge of steel), while the dimensional checks other than pipe end wall thickness and weight are made on a percentage basis of pipes produced per shift or per day. Corrosion testing is normally performed once per heat for the first three heats and repeated for every tenth heat.

The non-destructive testing of welds are made by either x-ray or ultrasonic testing, and supplemented by internal magnetic particle inspection. Some pipe production routes are also suitable for the use of eddy current testing. The test procedures are pre-approved and checked frequently during production, and the equipment, procedures and acceptance criteria are described in the quality assurance and quality control documentation.

The purpose of the hydrostatic testing (mill test) is to document the strength of the linepipe material. Thus the specified test pressure would typically be such as to subject the pipe wall to a hoop stress that is either 96% of the specified minimum yield stress (SMYS) or 84% of the specified minimum tensile strength (SMTS), whichever value is lower, based on a wall thickness that is the nominal value reduced by the fabrication tolerance.

7.3 Internal coating

7.3.1 *General*

Internal coating of the pipeline may be specified to prevent internal corrosion, to resist erosion, or to reduce the flow resistance. The latter is only suitable for pipelines transporting a gaseous medium, such as natural gas, where the throughput may be increased significantly (by approximately 10%) by drag-reducing coating.

For an internal coating to be effective against corrosion it is necessary that the field joints, i.e. the areas where the individual pipe joints are welded together, be internally coated as well, otherwise the girth welds will be particularly exposed to internal corrosion. The same applies to the internal surface of linepipe bends, flanges and other components. The most common internal anti-corrosion coating is fusion bonded epoxy (FBE).

A drag-reducing internal coating is normally applied in the form of liquid, two-component epoxy paint. Although it will not protect against corrosion in the long run the coating will prevent the formation of corrosion products in the pipe joints during storage, thereby greatly reducing the amount of debris to be removed from the pipeline during pre-commissioning. Hence internal coating may also be prescribed in cases where the increase in throughput is marginal. On the other hand, drag-reducing internal coating is not applied to girth welds, bends and components.

Internal coating of the pipe joints may be performed at the pipe mill or at a dedicated coating yard. In the latter case, internal coating is often the first operation, although internal painting, which does not require significant heating of the pipe, may be performed after the application of external coatings. Indeed, external anti-corrosion coatings that entail heating of the pipe, such as three-layer polyolefins, will normally have to be applied prior to the internal epoxy painting, in order not to damage the latter.

An international specification ISO 15741 for internal drag-reducing epoxy painting is under preparation. Until it is approved the draft version ISO/DIS 15741.2 may be used as basis for project-specific specifications.

The capacity of an internal coating production line is typically 30–40 pipe joints per hour for epoxy paint, somewhat less for FBE.

7.3.2 *Surface preparation*

The pipe joints are dried and preheated, normally by gas flame lances inserted into the bore, and blast cleaned to remove mill scale, corrosion products or foreign matter from the internal surface.

The relative humidity during blast cleaning should be below 85%, and the temperature of the steel surface should be at least 3°C above the dew point. At high air humidity the dew point may be quite high, and it is not unusual to specify temperatures of around 70°C before blast cleaning.

The preferred blasting medium is steel grit, and the cleaned surface should have a uniform grey appearance (surface quality Sa 2.5 according to ISO 8501-1). The specified surface profile depends upon the subsequent internal coating, typical values being a trough-to-peak distance of 50–100 micron for FBE and 25–75 micron for epoxy paint.

After blast cleaning a distance (50–100 mm, depending upon the coating) at each end is masked off to provide a coating cut-back for the field joint welding. Coating should follow immediately after blast cleaning, but a delay of up to four hours is normally considered safe.

7.3.3 *Fusion bonded epoxy (FBE)*

The general principles for chemical pre-treatment, electrostatic heating, powder application, curing, repair, inspection and testing of internal FBE are the same as for the corresponding external anti-corrosion coating (see Section 7.4.5), except that application is carried out from the inside, using lances.

For very small diameter pipe (less than 3″) airless spraying of the epoxy powder is impractical, and a fluidised bed method, including the application of a liquid primer, has been developed. A fluidised bed is a container where the epoxy powder is kept suspended by an upward directed airflow, and into which the heated pipe joint is immersed until it has acquired a coating of the specified thickness.

7.3.4 *Epoxy paint*

For internal coatings of gas pipelines it is important that no solid matter (e.g. $C_{18}H_{20}$) be precipitated by cooling gas through contact with the coating; documentation of this may be required. Likewise it must be ensured that no solvents from the coating are released into the hydrostatic testing water. It may be required that paint batch samples be kept for a specified period (say one year) for verification purposes.

Application of two-component epoxy paint is normally by means of spray nozzles mounted on a lance inserted into the rotating pipe. The specified dry film thickness (DFT) may vary, but less than 100 micron (nominal) is not recommended.

Repairs may be carried out within specified limits (say maximum area 1% of the surface, no more than ten repairs per pipe joint), using power brushing, pre-heating, and soft paint-brush application of the same coating material.

7.3.5 *Inspection and testing of epoxy paint*

Visual inspection
All pipes are inspected to verify that the coating is free from runs, sags or other imperfections.

Coating thickness
On a specified proportion (initially every pipe, but subsequently reduced to, say, 10%) the dry film thickness is measured at a number of locations (say four at each end of the pipe). The accepted tolerance is typically ±20 micron.

Surface roughness

The roughness (Ra) of the painted surface is measured at the same locations as the coating thickness, and shall typically not exceed 25 micron.

Laboratory testing

For each paint batch the mixing ratio is checked by rubbing a cured test sample with a cloth soaked in one of the solvents of the mix, only a slight coloration of the cloth being acceptable. Furthermore, the adhesion is checked by making a cross cut with a sharp knife on a plane sample, and by bending a sample over a mandrel (of 13 mm, say). In neither case should there be any flaking or loss of adhesion.

7.4   External anti-corrosion coating

7.4.1   *General*

Traditionally, hot applied enamel coatings have been the dominant option for marine pipelines, and asphalt enamel remains the preferred choice for pipes provided with concrete weight coating. However, more sophisticated three-layer polyolefin coatings are gaining ground, and are normally specified for pipes that are not mechanically protected by concrete. Fusion bonded epoxy (FBE) is not much used for marine pipelines as the coating is susceptible to mechanical damage during handling and transport, and does not resist concrete coating by impingement.

A typical exception is constituted by small diameter pipelines installed by reeling (see Section 8.4.4), and which during installation offshore will be connected to sacrificial anodes or to another cathodically protected pipeline. To protect the integrity of the linepipe steel such pipes are often provided with doubler plates, i.e. small patches of steel plate that are welded on to the pipe joint under controlled factory conditions, and on to which the electrical cables are welded offshore. FBE is very suitable for such irregular pipe, and the thin film coating is easily removed from the doubler plates for offshore welding.

External polyolefin coatings may be applied at the pipe mill as they are resistant to handling and transport damage, which is not the case for enamels and FBE. Normally, however, external coatings, particularly if concrete is included, are applied at a dedicated coating yard. At the coating yard the external coating normally follows immediately after the internal coating, unless the external coating process includes heating that would damage the internal coating (which is the case for external polyolefin coating combined with internal epoxy paint). Then pipes that are not (yet) internally coated will be brought in directly from the linepipe storage area. Some internally coated pipe joints may also be in intermediate storage because the production rate of the external coating typically is lower.

The pipe is normally coated as a continuous, rotating string, moving through the facility. The individual pipe joints are subsequently separated and provided with the appropriate coating cut-back, if this is not already achieved by masking tape or by coupling devices shielding the pipe ends. Linepipe bends, which cannot be rotated, are normally provided with hand-applied elastomer external coating or flame-sprayed polyolefin. Alternatively, an epoxy paint system may be applied, although it will entail a drain on the cathodic protection system due to the higher coating breakdown ratio; see Section 6.7.4.

Specifications for the several types of external coatings have been issued by a variety of organisations. A typical example, which covers a range of coatings, including hot applied enamels, multi-layer polyolefins, FBE and elastomers, is the DNV RP F106, issued by Det Norske Veritas.

7.4.2 *Surface preparation*

Prior to blast cleaning all salts or other contaminants are removed from the pipe surface, typically by fresh water washing with hot water or steam. Immediately before blast cleaning the external pipe surface is dried and preheated, the temperature depending upon the subsequent coating.

Tests for salt (NaCl) contamination may be specified, particularly for sensitive coatings such as FBE or elastomer, acceptable maximum levels being in the order of 20–40 mg/m^2, the former value being specified by DNV RP F106.

Using steel grit, the external pipe surface is blast cleaned to a uniform grey appearance (surface quality Sa 2.5 according to ISO 8501-1). The specified surface profile depends upon the subsequent external coating, a typical requirement being a trough-to-peak distance of 50–100 micron. Moreover, the anchor pattern shall have a high peak density (see ISO 4287/1 *Surface roughness and its parameters*). The blast-cleaned surface is vacuumed to remove any dust, and any steel surface defects are removed by light grinding. The pipe interior is emptied of abrasive or other foreign matter, in order not to contaminate the external coating, which should be applied within a specified time of blast cleaning, depending upon the type of coating.

The adhesion of the subsequent coating can be greatly enhanced by chemical pre-treatment, notably acid wash and chromate conversion coating. Acid washing usually uses diluted phosphoric acid, which effectively removes any damaging contaminant, such as inorganic salts, oil and grease, etc. The acid solution is flooded or sprayed on to the rotating pipe, which is heated to the temperature specified by the manufacturer. The pipe surface should be fully wetted in a revolution or two, and the chemical is allowed to remain for about 20 seconds, whereupon all traces are removed by high pressure clean water rinsing. Longer reaction times will result in a passive phosphate layer on the steel surface, visible as a light blue coloration or traces of white powder. The acid wash should be applied after grit blasting, or possibly between a first and a second blast cleaning, which will also remove any passive layer that may have formed. Obviously, extreme care should be taken that the surface is not contaminated by the blast medium.

Dry-in-place chromium/silica complex oxide conversion coating is a pretreatment process that involves the application of the unheated chromium treatment chemical to the rotating pipe surface. Spraying is normally not permitted for safety reasons, and typically rubber squeegees are used to wet the surface evenly within 1–2 revolutions of the pipe. The pipe then needs to be heated to above 120°C for at least 20 seconds (for FBE application see below, this is already a part of the coating process). The result is the formation of a 1–2 micron thick integral layer of mixed chromium, silicon and ferrous oxides, visible as light golden-bronze coloration of the steel surface. This facilitates the chemical bonding of coating materials, notably FBE, and increases the resistance to any corrosive ions.

7.4.3 *Asphalt enamel*

The hot applied coating consists of bitumen enamel reinforced with one or more layers of fibreglass tissue inner wrap, and provided with an outer wrap of bitumen impregnated fibreglass felt. Traditionally, the total thickness would be 5–6 mm (see Figure 7.1), but recent specifications allow 4 mm, provided that it is achieved over the weld seam as well. The hot enamel is prepared from oxidised bitumen mixed with mineral fillers to the specified hardness and softening point, and stored at a temperature of approximately 220°C.

For primer application the pipe temperature shall be approximately 40°C, if needed the pipe is heated from the inside by hot air. A synthetic primer is applied by airless spraying equipment to a dry film thickness of approximately 20 micron.

The hot enamel is applied when the primer has dried, but still remains 'tacky'. To prevent the formation of voids along the longitudinal weld seam it is recommended that one places a string of enamel along the weld before the pipe starts to rotate and enter the enamel coating booth; see Figure 7.9. In the coating booth the hot enamel is flooded on to the pipe, and at the same time the fibreglass inner wrap is spirally wound on to the pipe from one or two rolls of material; see Figure 7.10. Similarly, the impregnated outer wrap is applied after the enamel flooding weir. Finally, the pipe joints are separated, and the coating is quenched by water, see Figure 7.11.

After cooling to approximately 80°C the coating at the pipe ends is removed to create an appropriate cut-back, typically 250–280 mm, and the coating is chamfered at 45°. Cleaning of the pipe ends may be specified, but the primer may as well be left on as a temporary corrosion protection during subsequent storage, except for the bevels and adjacent area (say 50 mm), where it could give rise to noxious fumes during field joint welding.

Figure 7.9   Enamel applied to weld seam of primed pipe

Figure 7.10   Enamel inner wrap and outer wrap applied to pipe joint

Figure 7.11 Water quenching of enamel coated pipe joint

Repair of pinholes may be carried out by the application of hot enamel only. Larger areas should be cleaned and power brushed, the surrounding coating chamfered, and liquid primer applied. Enamel, inner and outer wraps are then used to build up the coating to the original thickness.

The temperature tolerance of asphalt enamel depends on the grade of bitumen used, but the coating would not be recommended for service temperatures exceeding 70°C, and for pipelines where interface shear is expected a limit of 60°C should prudently be adopted. By use of so-called modified asphalt enamel, which is produced by mixing bitumen with elastomer or rubber, the temperature range may be extended, reportedly up to 80°C.

7.4.4 *Three-layer polyolefin coatings (PE/PU/PP)*

Modern three-layer coatings consist of an epoxy primer, a copolymer adhesive, and a top layer of polyethylene (PE), polyurethane (PU) or polypropylene (PP). The main difference between the three options is increasing temperature toler-ance, which is matched by increasing price levels. Thus PE should not be used for service temperatures exceeding 85°C, whereas up to 100°C can be permitted for PU systems, and PP coatings perform satisfactorily between 75°C and 140°C. However, these temperature limits are indicative only, as polyolefin coatings are constantly being developed by manufacturers.

The superior performance of the three-layer coatings is due to the fact that the copolymer adhesive forms a chemical bond with epoxy, but not with steel, whereas epoxy bonds very well to steel, particularly after a chemical pre-treatment using a phosphoric acid or chromate wash of the substrate, see Section 7.4.2.

The pipe is electrostatically heated to approximately 220°C before entering the FBE coating booth. The epoxy powder is applied by airless guns, and electrostatically attracted to the pipe, where it sinters together on the hot surface. Overspray powder is recycled into the system at a specified maximum ratio (say 10%). The epoxy primer thickness is a minimum of 50 micron, and superior performance is obtained with a thickness of approximately 150 micron (over the shot blast anchor pattern).

When the epoxy is gelling, but has not completely cured, the copolymer adhesive is spray applied to a thickness of 140–200 micron, no copolymer reclaim being used. Recently an adhesive of polyethylene grafted with maleic anhydride molecules has been developed, with improved bonding strength to the fusion bonded epoxy. Recent investigations have also shown that equivalent coating properties may be obtained by replacing the fusion bonded epoxy primer with a liquid, solvent-free epoxy.

Immediately afterwards the polyolefin top layer is side extruded onto the pipe, using one or several extruder heads. The soft polyolefin sheets are smoothed with rollers to form a seamless coating of the specified thickness, normally 2–3 mm, which may be reduced to 1–2 mm if the pipe is to be concrete coated. For smaller diameter pipe (up to 12″ at least) annular (crosshead) extrusion is often used instead of the side extruders. The pipe joints are separated and water quenched to a temperature below 80°C for inspection and chamfering of the coating at the cut-back.

Repairs to the top layer only may be carried out using extruded polyolofin as used in the parent coating. For repairs involving the entire three-layer system a suitable area (say 50 mm diameter) must first be cleaned and power brushed, and a two-component epoxy compound applied. Normally a maximum repair area (say 40 cm^2) is specified.

7.4.5 *Fusion bonded epoxy (FBE)*

FBE as an anti-corrosion coating in its own right predates its use in three-layer coatings. The application is as described above, but to be effective as corrosion protection the FBE dry film thickness should be at least 400 micron. The high temperature tolerance of the FBE in three-layer coatings is due to the fact that it is essentially dry. As a stand-alone coating in contact with water, FBE should not be used at service temperatures exceeding 70°C.

The thin film coating is vulnerable to mechanical impact, and an unacceptable number of holes is likely to occur during handling and transport, unless the coating is protected with an overlay, e.g. of fibre reinforced cement mortar or liquid applied polymer concrete. A limited number of pinholes may be repaired by epoxy melt sticks, see Figure 7.12.

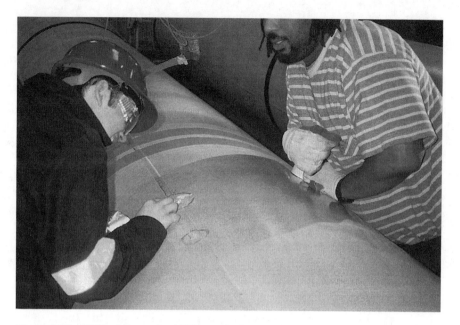

Figure 7.12   Melt stick repair of FBE coating defect

Mechanical protection of FBE is necessary if concrete coating is applied by impingement, whereas extrusion or slipforming (see Section 7.7) are feasible options. In all cases, however, measures must be introduced to prevent longitudinal slippage of the concrete over the smooth FBE coating. This may be achieved by additional epoxy powder applied in a spiral pattern or by a suitable topcoat, see below. For the above reasons, FBE coating is not much used for marine pipelines, except in the USA, where FBE is a very common pipeline coating on onshore pipelines, and therefore quite competitive.

A recent development is the introduction of dual powder FBE coatings, which feature a base layer FBE with a topcoat of modified FBE to give resistance to damage. To provide a chemically bonded coating the top layer is sprayed onto the base layer before the latter has fully cured. The total dry film thickness is in the range 1.0–1.3 mm, and the performance is comparable to three layer polyolefin coatings. To prevent concrete coating slippage the topcoat may be formulated with a sandpaper-like texture, which also improves the adhesion of the concrete mix during impingement. The product can be applied over a dual powder system or it may replace the second layer.

7.4.6   *Elastomer coating (Neoprene)*

Neoprene is a thick (typically 12–15 mm) coating made from unhardened polychloroprene elastomer sheets, which are wrapped around the pipe, and subsequently vulcanised in an autoclave. The sheets are normally hand-applied to a stationary pipe, which makes the coating suitable for linepipe bends.

Within two hours after blast cleaning and testing for salt contamination a primer is brush applied to the pre-heated pipe, followed by a bonding agent. The elastomer sheets are milled and cut to the required thickness and suitable dimensions, and stored at the manufacturer's recommended temperature. Before application to the pipe the sheets are solvent washed to render them 'tacky'. The edges are chamfered and/or stitched together to form a smooth coating. A maximum time lapse (say 96 hours) should pass between the application of primer and bonding agent, and of bonding agent and elastomer.

The coated pipe is tightly wrapped with nylon tape and placed in an autoclave for vulcanisation. The curing should start within a certain time after elastomer application, typically 72 hours, less in case of high ambient temperatures. The curing cycle involves heating and pressurisation (say to 145°C and 500 kPa) over at least 2 hours. After curing and cooling the nylon wrap is removed.

Repairs to neoprene coatings may be carried out using elastomer, bonding agent and primer (if the defect extends to the steel surface). The repair shall be cured in an autoclave as the parent coating, or alternatively by hot tapes wrapped around the repaired area. Minor surface defects may be repaired by two-component epoxy mixed with cured elastomer powder, left overnight to cure naturally.

7.4.7   *Thermal insulation*

Thermal insulation may be incorporated into the external anti-corrosion coating. The procedure is the same as for the three-layer polyolefin coatings described in Section 7.4.4 above, except that the outer layer is replaced by a foam, typically based on polypropylene (PP), polyurethane (PU) or PVC, which is extruded, flooded or sprayed onto the pipe. As foaming agents fluorocarbons have been widely used, but for environmental reasons they are being replaced by more benign compounds, such as CO_2.

The foam may be resistant to water ingress, or an outer solid layer (e.g. solid PU over PU foam) may be added. In that case the solid skin is tapered down to the FBE coating at the ends, making the pipe joint similar to a pipe-in-pipe system; see below. Note that if the pipeline is to be installed by laybarge the coating must resist not only the external water pressure, but also the squeeze pressure from the laybarge tensioners; see Section 8.4.2.

The insulation properties of a foam depend upon the density, and denser foams must be used to resist the external pressure at increasing water depths. For large water depths it is possible to use syntactic foam, where the air voids are replaced by hollow glass microspheres, which are incorporated into the polyolefin as it is extruded. Syntactic foams are hydrophobic, which means that no external barrier is needed, and it is possible to reach water depths exceeding

2000 m. For greater depths it is necessary to use solid polyolefin throughout. The thermal conductivities of coating materials range from 0.04 W/m/°C (PU foam) through 0.1–0.2 W/m/°C (syntactic foams) to 0.4 W/m/°C (HDPU); see Section 3.5.3.

An alternative approach to insulation is to use a pipe-in-pipe system, where the product pipe is inserted into a larger sleeve pipe that resists the water pressure, and the annulus is filled with insulating foam, typically PU. The filling may be carried out by placing the mixture of polyolefin and foaming agent on a tape, which is pulled into the sleeve pipe, once the product pipe is in place. Alternatively, the annulus may be filled with alumina silicate microspheres. The sleeve pipe may be made from steel or high density polyolefin (HDPE or HDPU), depending upon the water depth. At the ends of the pipe joint the annulus is closed off, typically by heat shrink sleeves with a specially high shrink ratio. For steel sleeve pipes welding may be used, and proprietary steel-in-steel systems have been developed where the product pipe and the sleeve pipe are welded on to specially designed end pieces.

It has been discussed whether such fixed, shear-transferring connections (bulkheads) are needed along the pipeline to prevent excessive temperature movements of the product pipe relative to the sleeve pipe, but it may be shown that they are not necessary, except at the ends of the pipeline. On the other hand, it is prudent to compartmentalise the pipeline, so that accidental water ingress into one pipe joint (or field joint) does not compromise the insulation of the entire pipeline. However, these seals, like the heat shrink sleeves mentioned above, need not be able to transfer shear between the two pipes. For installation by laybarge it is necessary that the lay tension can be transferred from the sleeve pipe to the product pipe through the insulation, and if this is not the case shear connectors must be provided.

Steel sleeve pipes are provided with anti-corrosion coating like any other pipeline, and cathodic protection is achieved by conventional sacrificial anodes, attached to the sleeve pipe as described in Section 7.6 below. However, in the case of a polyolefin sleeve pipe the electrical connection must be made to the product pipe at the field joint area at the end of the pipe joint, whereas for a steel-in-steel system no cathodic protection of the product pipe is needed.

7.4.8 Inspection and testing

Visual inspection
All pipes are inspected to verify that the coating is free from defects or imperfections.

Holiday detection
The entire coating is examined for pinholes by checking the electrical insulation. A voltage depending of the type of coating is applied to the pipe, and an annular

Figure 7.13   Hole detection on FBE coated pipe joint

wire coil is run along the pipe joint; see Figure 7.13. Any pinhole will generate a spark, which is detected by amplification of the sound. The holiday detector is routinely calibrated, typically against an artificial pinhole.

Coating thickness

On a specified proportion (initially every pipe, but subsequently reduced to, say, 10%) the coating thickness is measured at a number of locations (say four at each third of the pipe length), typically by a magnetic or electromagnetic thickness gauge with an accuracy of ±10%.

Adhesion

The adhesion of the coating to the steel substrate is tested at the beginning of a shift, or after major shut-downs, typically by pulling off a strip separated by knife incisions. For the thick coatings the absence of delaminations is also checked by light tapping along the pipe joint.

Cathodic disbonding

The alkalinity associated with cathodic protection may reduce the adhesion of the coating, a phenomenon known as cathodic disbonding. FBE in particular is susceptible, and documentation of the properties is therefore often required for FBE or three-layer polyolefin coatings. A test sample is provided with an artificial hole, placed in a specified solution, and subjected to a specified voltage for a specified period. A typical test standard is ASTM G8.

Laboratory testing

Depending upon the type of coating, specific laboratory tests may be required, such as tests for:

- bacterial resistance;
- water absorption;
- thermal conductivity;
- density;
- modulus of elasticity, tensile strength and linear expansion (polyolefins, polychloroprene);
- softening point and penetration (enamels).

7.4.9   *Other coating systems*

Tape wrapping

Cold applied tape for corrosion protection of pipes is a two-layer system, consisting of a polymer (typically PVC or PE) carrier tape, with an adhesive (typically bitumen or butyl rubber). The tape is spirally wrapped under tension on to the rotating pipe with or without prior application of primer. The amount of overlap determines the total coating thickness, which is typically 2 mm; if needed, several layers can be applied.

Cold tape wrapping is relatively tolerant with respect to surface preparation, and wire brushing may be substituted for blast cleaning. The system is widely used for onshore pipelines, where the coating is applied 'over the ditch' by wrapping machines spiralling around the stationary pipe, but the quality is generally not considered suitable for marine pipelines, where the costs of coating repairs are orders of magnitude higher. An exception is the coating of field joints; see Section 8.3.3.

Somastic coating

Somastic is an extrudable mastic consisting of oxidized bitumen mixed with mineral filler materials. The hot mastic is spirally extruded onto the pipe, forming a seamless coating of thickness approximately 15 mm. The mechanical strength of the coating is very low, and the system is hardly used any more, being replaced by the hot applied enamels.

Coal tar enamel

Hot applied coal tar enamel coating is identical to the corresponding asphalt enamel, except that the bitumen is replaced by coal tar, distilled by pyrolysis of rock coal. Formerly, coal tar enamel coating was preferred because of its resistance to higher temperatures (up to 85°C), and because coal tar is toxic to organic growth. The very toxicity, however, raises concerns over environmental impact, as well as workers' health and safety during coating application, and consequently the use of coal tar enamel has been prohibited in most countries.

Sintered polyethylene (PE)

The application of sintered PE is similar to that of FBE, in the sense that the PE powder is electrostatically sprayed onto the hot pipe, where it melts to form a seamless coating. A typical thickness is 4 mm, and a certain volume of linepipe steel is necessary to provide the heat capacity to melt the corresponding layer of PE, hence the system is not suitable for thin-walled pipe. Most marine pipelines would have sufficient wall thickness, but sintered PE coatings are not much used any more.

Two-layer polyethylene/polypropylene (PE/PP)

Two-layer polyolefin coatings are applied by extruding a copolymer adhesive onto the pipe, followed by a top layer of PE or PP, the main difference between the two being that PP is suitable for higher temperatures. For small diameter pipes the extrusion may be longitudinal, using an annular crosshead extruder, but for the pipe sizes suitable for marine pipelines it is normal to apply side extrusion onto the rotating pipe. The soft polymer sheets are smoothed with rollers to form a seamless coating of thickness 2–3 mm. Two-layer polyolefin coatings are not much in use any more, being replaced by the high performing three-layer coatings.

Solvent free phenolic epoxy

Solvent free phenolic epoxy can be spray applied in a thickness of up to 1.2 mm in one application. The performance is similar to that of FBE and three-layer coatings, except that the temperature tolerance is as high as 160°C. The coating can be applied to substrates with temperatures up to 90°C, and is therefore mainly used for landline rehabilitation. The application for marine pipelines might be suitable in cases where the operational temperature over a short distance exceeds the tolerance of more conventional coatings.

Metal sheeting

External liners (sheetings) are manufactured from corrosion resistant alloys, e.g. Monel or 90/10 Cu-Ni, and they are used only at very exposed locations, such as the splash zone of pipeline risers. All connections in the sheeting or to the substrate should be welded, and subjected to 100% tightness inspection. It should be noted that the sheeting to some extent prevents condition assessment of the underlying linepipe.

7.5   Anode manufacture

7.5.1   *General*

Sacrificial anodes for marine pipelines are normally of the bracelet type, i.e. a cylindrical shell around the pipe joint, and are produced by specialist anode manufacturers to specifications satisfying the requirements of Section 3.4. The usual materials are alloys based on zinc or aluminium, the latter becoming most prevalent for environmental as well as economic reasons.

It is possible to cast zinc anodes directly on to steel pipes, but pipeline anodes are normally manufactured as two half-shells, which are clamped around the pipe on top of the anti-corrosion coating. For practical reasons related to casting, the anode length should not exceed 1.0 m, and its minimum thickness is 40 mm.

When the anodes are installed on individual pipe joints in a coating yard the mounting involves welding of protruding reinforcement straps on both sides; see Section 7.6.2. Anodes mounted offshore on a reeled pipeline are normally bolted together; see Figure 7.14 and, in the absence of a concrete coating, the anode is tapered at the ends to facilitate installation. A typical bracelet anode mounted on a concrete coated pipe is shown in Figure 7.15. Note also the anode joints being unloaded from the pipe supply vessel in Figure 8.9.

Figure 7.14   Aluminium anode for offshore installation

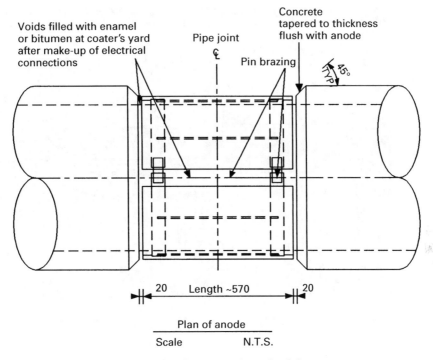

Voids filled with enamel
or bitumen at coater's yard
after make-up of electrical
connections

Pipe joint
₵

Concrete
tapered to thickness
flush with anode

Pin brazing

45° TYP

20   Length ~570   20

Plan of anode

Scale   N.T.S.

Figure 7.15 **Typical bracelet anode mounted on pipe joint**

7.5.2 *Reinforcement insert*

A steel reinforcement is introduced in each anode half-shell to provide mechanical strength and integrity as the anode is being consumed. The insert is welded from a number (typically two) of circumferential flatiron straps and a number (typically three to five per half-shell) of longitudinal bars. The flat straps protrude from the straight faces of the half-shell, and connect the two halves when the anode is installed. They are also used for attachment of the electrical cable connection; see Figure 7.19 below, as well as for lifting and handling the half-shell. The latter may be facilitated by providing protruding lugs with holes for the attachment of hooks; see Figure 7.17. Figure 7.16 shows batches of reinforcement inserts.

The insert, which is fabricated from mild, weldable steel, should be blast cleaned a specified maximum time (typically 12 hours) before the anode half-shell is cast. To ensure the integrity of the anode at the end of its life the reinforcement should be far from the outer anode surface; one possibility is to place the circumferential straps flush with the inner surface. For aluminium anodes this is preferable because the lack of chemical bond to steel implies that aluminium that is located under the insert strip is not accessible to provide protection.

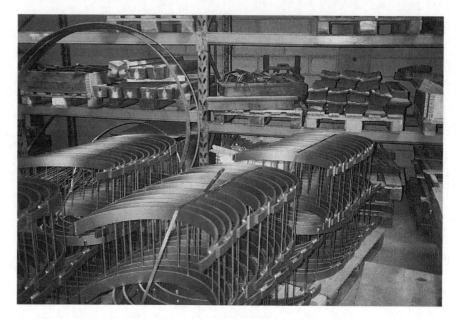

Figure 7.16   Reinforcement inserts for sacrificial anode half shells

7.5.3   *Alloy materials*

The materials normally used for pipeline anodes are zinc or aluminium alloys.

Zinc anode alloy is traditionally based upon the US Military Specification A-18001 H, and typical material compositions are given in Section 3.4.2. Requirements should be specified to the electrochemical capacity and the closed circuit potential. Typical values (NORSOK Standard M-503) are 780 A h/kg, respectively 1.03 V (with respect to a Ag/AgCl/Seawater reference electrode).

Aluminium anodes are normally activated by a small amount of indium, and typical material compositions are given in Section 3.4.3. Requirements should be specified to the electrochemical capacity and the closed circuit potential. Typical values (NORSOK Standard M-503) are 2500 A h/kg, respectively 1.07 V (with respect to a Ag/AgCl/Seawater reference electrode).

7.5.4   *Casting*

It is essential to avoid contamination of the alloy during preparation of the melt. For aluminium anodes the base metal ingots may be lowered into the furnace using straps made of aluminium.

The moulds for the half-shells may be cast in steel or welded from steel plates. The former are more expensive, but tend to give better results. To allow the alloy to flow around the insert the half-shells are normally cast lying down, i.e. with the cylindrical axis horizontal and the curved outer surface downwards. The

Figure 7.17   Sacrificial anode half-shells ready for installation

insert is strapped at the correct position on to the internal mould, which is then lowered into the external mould, and the metal is poured into the void through the straight anode faces with the protruding straps. Individual anode manufacturers have developed proprietary procedures of preheating, pouring and cooling to minimise cracking and suction of the metal.

As soon as the half-shells are cool enough to handle the shape is checked with templates and, if needed, adjusted with mechanical tools to correct any distortion outside the specified range. Any protrusions on the inner surface that could damage the anti-corrosion coating during anode installation are removed. Completed anode half-shells are shown in Figure 7.17.

7.5.5   *Anode defects*

Even with the best foundry practice it is difficult to avoid cracking, particularly of thin anodes cast from aluminium alloys. To ensure the integrity of the anode during its entire service life no cracks should be allowed in parts that are not

Figure 7.18   Acceptable surface crack in an aluminium anode

supported by the reinforcement insert but, apart from that, specifications vary somewhat with regards to permissible cracking. There is general agreement that the following can be allowed:

- longitudinal cracks of width ≤0.5 mm; length ≤20% of the anode length or 100 mm, whichever is less;
- circumferential cracks of width ≤0.5 mm; length ≤50% of the internal anode diameter or 100 mm, whichever is less.

Whichever criteria are adopted they should be pragmatic, reflecting the fact that the visual appearance of the anode is without consequence for its function. When setting a limitation on the permissible number of cracks it is customary to consider small and closely spaced cracks (less than 0.5 mm wide and separated by less than 3 mm) as one crack. Figure 7.18 shows a typical crack in an aluminium anode that should be considered acceptable.

In addition to cracking, limitations are given to other surface defects, such as shrinkage depressions, cold shuts or surface laps. Defects that expose the insert from the outer surface should not be allowed. The most critical internal defect is lack of contact between anode metal and insert, particularly for aluminium alloys, that do not form a chemical bond to the reinforcing insert steel. NACE RP 0492 specifies upper limits for internal defects, measured on cut surfaces.

7.5.6 *Electrical connections*

The cables providing the electrical connection to the pipeline should have a low resistance; normally 16 mm^2 single core copper with double PVC insulation is prescribed. The cable end may be thermite welded directly on a protruding insert strap, or a steel lug is attached to the cable end by thermite welding or mechanical crimping. Crimping is done by inserting the cable end into a short length of tightly fitting steel tube, which is then hydraulically squeezed around the cable. The other end of the tube is flattened for pin brazing or stick welding to the insert; see Figure 7.19. The free end of each cable is left bare for thermite welding to the linepipe steel, or provided with an end lug for pin brazing or stick welding on to doubler plates (anode pads); see Section 7.6.2, where thermite welding and pin brazing are described.

One cable is attached to each anode half-shell, thus providing double connection of the installed anode. The length of each cable should be sufficient to allow minor adjustments of the anode relative to the pipe, and yet be no longer than can be fitted between the two anode halves. For anodes to be mounted on a reeled pipeline (see Section 8.4.4), where the cable is not protected between the anode halves, more mechanically resistant steel cables (lightning rod types) may be used, and the cable length shall take account of the location of doubler plates welded on to the pipeline.

Figure 7.19   Anode cable with cable shoe for pin brazing on to the pipe

7.5.7 *Painting and marking*

To protect the inner anode surface it is customary to prescribe painting by coal tar epoxy or similar to a specified dry film thickness (say 100 micron). Obviously no paint should be allowed on the outer surface.

Each half-shell should be die stamped with the individual half-shell number, as well as the number of the melt from which it was cast. Additional markings may identify the supplier, anode type, inspection status, etc. To allow identification after the anode has been installed on the pipe joint, the number should be on the outer, curved surface.

7.5.8 *Inspection and testing*

Insert fabrication
Normally all welds are visually inspected, and magnetic particle inspection (MPI) of some (say 10%) of the welds may be prescribed.

Alloy composition
The composition of the alloy is checked against the specification, normally by means of spectrographic examination of a sample taken from the melt.

Electrochemical performance
The electrochemical properties of the anode material (capacity, closed circuit potential) are documented by laboratory testing. The capacity is determined by measuring the weight loss of a sample immersed in a test solution and subjected to specified current densities during a specified period, e.g. as prescribed by the DNV RP B401. The normal duration of the test is 96 hours, but long term testing in special environments, such as hot seawater or marine sediments, may also be prescribed.

Anode weight
As the lifetime of the anode depends upon the mass of anode alloy, strict tolerances must be placed upon the net weight, i.e. the weight of the half-shell minus the weight of the steel insert, the latter being determined by weighing a number (say 50) units. Typical tolerances are ±3% on the individual half-shell, and −0%, +2% on the total order.

Visual inspection
It is checked that the physical dimensions – diameter, thickness, length, insert position, etc. – as well as the shape, are within the prescribed tolerances. It is customary to specify the internal diameter and the shell thickness, and the supplier then determines the length and the gap between the half-shells to comply with the specified weight. All half-shells are inspected to verify compliance with the acceptance criteria for cracking and other surface defects.

Internal examination

A specified proportion (say 2%) of the half-shells are sectioned transversely (typically by four cuts) to verify compliance with the acceptance criteria for internal defects and position of the insert.

Cable connection testing

Each cable connection is tested for mechanical strength by pulling in the longitudinal direction, the normal criterion being that the connection shall resist 'a full body weight'. A specified proportion (say 2%) are tested for electrical continuity, the normal criterion being that the resistance shall not exceed 0.01 Ω.

7.6 Anode installation

7.6.1 *General*

Sacrificial anodes are mounted on to pipe joints with or without concrete coating and in the former case the operation normally follows immediately prior to the application of concrete. An anode bracelet is formed by joining the steel inserts of the two half-shells, and electrical cable connections are made to the linepipe steel. The installation is normally carried out according to a project-specific specification. To minimise the installation time, whether onshore or offshore, two anodes may be installed adjacent to one other on the same pipe joint (twin anode).

7.6.2 *Anode mounting*

The two anode half-shells are lifted on to the externally coated pipe joint and fitted tightly by straps and come-alongs. For practical reasons the same nominal size anodes are often attached to pipe joints of slightly varying external diameters. This is because the different zoning along the pipeline may result in different wall thicknesses, and a constant bore is desirable to facilitate internal pigging. Since closing is easier than opening, the anodes are sized to give a tight fit to the largest external diameter. Final adjustment is carried out by brute force; see Figure 7.20.

When the anode is firmly positioned the overlapping insert lugs are welded together, the underlying anti-corrosion coating being protected against heat and weld spatter by heat-resistant sheet material. The location of the anode is normally between 3 m and 4.5 m from the end of the joint to minimise interference with the rollers supporting the pipe during double-jointing on the laybarge, and the gaps between the anode halves are positioned away from the longitudinal weld seam.

Anodes for installation on the pipe string during reeling are normally bolted together rather than welded (see Figure 7.14) and the nuts are secured by tack welds. To minimise offshore work, such anodes may be manufactured with hinges at one side, but this can make it more difficult to achieve a tight fit to the

Figure 7.20   Mounting of sacrificial anode on enamel coated pipe joint

pipe. Sliding of the anode along the pipe may result from pipelaying, trenching or soil resistance to temperature expansions during operation. This can be prevented by the provision of friction collars or other means, such as connecting an anode on each side of a raised field joint.

7.6.3   *Electrical connection*

In each of the gaps between the anode halves the anti-corrosion coating is removed from an area, and the bare metal exposed. Each anode half is equipped with an electrical copper cable welded to one insert lug, with the other end free for attachment to the cleaned pipe steel by pin brazing or thermite welding.

For thermite welding, the bare cable end is inserted into a mould that is filled with exothermal metal powder, placed on the pipe steel and ignited. The resulting metal lump may penetrate the pipe steel; procedure qualification may be required to document that this will not compromise the ductility of the pipe wall.

For pin brazing, the cable end is fitted with a metal lug with a hole through which a pin is brazed on to the pipe steel, see Figure 7.19. The brazing pin is made of brass with braze metal and flux, and the solder heat is much less than that for thermite welding.

Even pin brazing is not advisable on corrosion resistant alloy linepipe, and such pipelines are therefore usually provided with factory welded doubler plates (anode pads), on to which the electrical connection is made. However, the anode pads cause stress concentrations, which may lead to cracking during service. The problem is exacerbated by hydrogen embrittlement caused by the cathodic protection (see Section 6.7.2), particularly at the anode pads, where the potential is lowest and the coating most likely to be defective. One solution is to avoid bracelet anodes on CRA pipe; instead the anodes may be deployed as anode assemblies placed on structures electrically connected to the pipeline, or on special anode joints made from clad pipe. The reach of an anode assembly is a couple of kilometres, depending upon the attenuation through the pipeline, making the concept feasible for the relatively short flowlines that are most likely to be made from CRA.

It is not normal to weld anode pads on carbon steel linepipe, except for reeled pipelines, where the offshore electrical connection can then be made by stick welding of steel cables, adding to the shear resistance of the anode; Section 7.6.2 above.

7.6.4 *Finishing*

After completion of the electrical connection the anti-corrosion coating is restored, using the appropriate repair procedures, and the gap between the anode halves is filled with mastic or other suitable material. The external surface of the anode should be kept clear of all coating material.

7.7 Concrete coating

7.7.1 *General*

For some of the earlier marine pipelines the concrete weight coating was manufactured in the traditional way, by casting in shutters around the pipe, and this method may still be used for site application of concrete coatings of minor stream crossings, etc. For long transmission lines, however, this procedure is too costly and time consuming, and methods have been developed that dispense with the formwork.

The most common method of application of concrete is by impingement, a process whereby a fairly dry (no-slump) concrete mix is thrown at a rotating pipe. Other methods include wrapping concrete around the pipe, the mix being supported by a carrier tape, or casting in a sliding mould (slip-forming). To leave room for the girth welding equipment at pipe assembly the concrete coating is stopped at a distance (typically 360 mm) from the pipe ends. Reinforcing steel is provided, either in the form of prefabricated cages or as welded wire mesh, which is wound around the pipe simultaneously with the concrete application. The concrete coating is normally carried out in accordance with dedicated specifications, some guidelines being provided by, for example, DNV OS-F101.

The primary function of the concrete coating is to provide negative buoyancy to the pipeline, and the concrete density is increased by the addition of iron ore aggregate. Traditionally a density of 3040 kg/m^3 (compared with a normal concrete density of 2400 kg/m^3) is specified, but densities of 3300–3400 kg/m^3 can routinely be achieved, and up to 3800 kg/m^3 is possible with the slip-forming method.

7.7.2   *Concrete mix design*

The constituent materials are normally required to comply with usual standards for concrete production, e.g. EN 206-1. Depending upon the location of the project, it is customary to require the use of low-alkali, sulphate resistant cement, or alternatively blast furnace slag cement, which has a good track record for marine applications. Upper limits to the water/cement ratio (say 0.40) may be given. A minimum cement content of 300 kg/m^3 is adequate, but many operators will use much more to achieve desirable mix properties.

7.7.3   *Reinforcement*

The reinforcement may consist of wire mesh, which is spirally wrapped around the pipe simultaneously with the application of the concrete. Standard 17 gauge reinforcing net may be user in sheltered waters, but the performance of such coatings has been less than satisfactory in more harsh marine environments, and better results are obtained with heavy-gauge welded wire mesh. Alternatively, high quality coatings are reinforced with re-bar cages, which may be bent from standard welded mats or tailor-made as spot-welded spiral cages. The reinforcement cages are placed on the pipe joint prior to concrete application, and held in position by spacers; see Figure 7.21.

The reinforcement ratio should be sufficient to ensure the integrity of the coating during handling, transport, installation and operation, including the impact of fishing equipment. DNV OS-F101 specifies a minimum of 0.08% in the longitudinal direction, and 0.5% circumferentially. Electrical contact to the pipe steel or to any anodes must be avoided.

7.7.4   *Impingement*

The most common method of concrete application is by impingement. The concrete from the mixer is delivered on to a conveyor belt that transports it to the impingement unit, where fast rotating wire brushes or paddle arms literally throw it at the rotating pipe, which is travelling past the unit; see Figure 7.22.

Figure 7.21   Spiral reinforcement cages mounted on an enamel coated pipe joint

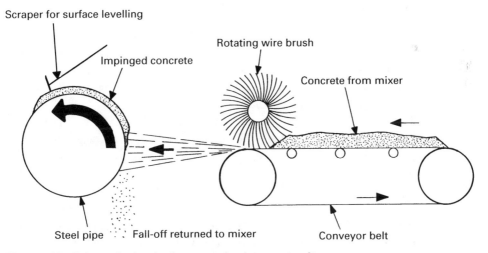

Figure 7.22   Schematic sketch of concrete impingement unit

The reinforcement may be pre-installed cages or wire mesh wound around the pipe during impingement. By adjusting the travel speed and the rate of rotation of the pipe the operator obtains the desired thickness of coating, which is then smoothed by scrapers. It is particularly important to avoid bulging of the coating at the pipe ends, as this can lead to spalling and radial cracking when the coated pipes are stacked. For the same reason the coating edges are slightly rounded at the ends, which are otherwise finished off square.

A certain amount of rebound material (say 10%) may be recycled into the mix, but to preclude excessive fall-off the impingement method requires a very dry (no-slump) concrete, with the consistency of moist earth. No large particles must be present (maximum aggregate size approximately 10 mm), and a large amount of fines are included. Thus a cement content of up to 600 kg/m^3 is not uncommon, resulting in water/cement ratios of about 0.30.

For practical reasons the minimum concrete thickness is 40 mm, and up to approximately 150 mm can be applied. In principle greater coating thickness can be obtained by applying a second layer after the curing of the first. When the specified thickness is reached, experience shows that a single steel wire wound around the pipe reduces the propensity for fall-off. However, this wire should be removed after curing of the pipe joint so that it doesn't interfere with the pipelaying equipment. Alternatively, filament strings may be incorporated into the coating, see Figure 7.26. Repairs to the uncured coating may be performed by shotcreting or trowelling of a slightly wetter mix.

The productivity of the impingement method is impressive, a well-functioning unit being capable of coating 1–1.5 km of pipe per 8 hour shift, corresponding to the application of approximately 300 m^3 of concrete.

7.7.5 *Extrusion*

In this method the concrete mix is extruded on to a polyethylene carrier tape, which is then spirally wrapped around the rotating pipe under compression. Reinforcement is fed into the coating in the form of wire mesh and/or spiral reinforcing bars; reinforcement cages cannot very well be applied. The carrier tape is normally removed after curing of the concrete.

The coating applied by extrusion can have any thicknesses ranging from 25 to 175 mm, and the productivity is of the same order of magnitude as for impingement.

7.7.6 *Slipforming*

The use of traditional concrete casting has been revived, using slipforming techniques. The individual pipe joint is provided with a reinforcement cage, placed in an upright position, and the concrete mix is poured into an annular slipform, which travels up the pipe under vibration. By mechanisation of the process, the capacity per unit can be increased to 4–5 joints per hour, corresponding to 0.4–0.5 km per 8 hour shift.

7.7.7 *Concrete curing*

To obtain a reasonable hydration of the cement it is important to retain what little moisture there is in the mix. This may be achieved by plastic wrapping or steam curing. In the latter case the temperature should be rather low (30–40°C), partly to avoid damage to the anti-corrosion coating (where asphalt enamel or similar is used) and partly to avoid dehydration of the concrete. After steam curing for 6–8 hours the concrete coated joints can be stacked for storage, but should be kept moist for 4 days.

Irrespective of curing method, the temperature should not fall below 5°C until the strength has reached 15 MPa. No pipe should be loaded out before a minimum of 7 days, and usually the pipes are kept longer to document the strength at 28 days; see below.

7.7.8 *Inspection and testing*

Strength

The determination of the compressive concrete strength represents a difficulty. Conventionally, structural concrete strength is specified as the compressive strength of standard 300 mm by 150 mm cylinders or 100 mm cubes, tested at 28 days, and such specimens are routinely cast and tested to monitor the quality of the concrete coating. However, due to the method of application and the dryness of the mix (particularly for impingement) the values may not be representative for the finished coating, and in situ testing is required to document the strength of the concrete actually applied on the pipe. Pipe coaters traditionally use drilled-out cores for this purpose, and DNV OS-F101 requires a minimum core strength of 40 MPa, determined according to the standard ASTM C39.

To give reasonably consistent results the core diameter should be at least three times the maximum aggregate size, and the length should be greater than the diameter. Coring should not damage the anti-corrosion coating, nor the reinforcement. It is quite difficult to extract undisturbed cores, particularly if wire mesh is used, the result being a large scatter of the results. Alternative in situ testing, such as pull-out tests, may be used. The best way to document the concrete strength is by specifying a characteristic strength on cylinders or cubes, and requiring the coater – by means of a test series – to establish a consistent relationship between this strength and the characteristic strength determined by the preferred in situ test method applied to the same specimens. The acceptance may be based upon the results at 7 days, provided a similar relationship is established between the 7-day and the 28-day strengths.

Density

The stresses in the pipe steel during installation are quite sensitive to the submerged weight of the pipe, therefore narrow tolerances are necessary on the coated pipe weight, typically −10% to +20% on an individual pipe joint and 0% to +4% on a day's production. To achieve such accuracy it is necessary to maintain a uniform concrete density, and samples are continuously prepared for the determination of dry density. During production a relationship is established between the dry concrete density and the density of the wet mix, allowing monitoring to be based upon sampling at the concrete plant.

The submerged weight of the pipe is also affected by the water absorption. For design purposes, the water absorption is normally taken at 2% (by weight), but it may well be higher, thus DNV OS-F101 allows up to 8% (by volume). At any rate, the water absorption is verified at the coating yard by the immersion of samples (or complete coated pipe joints) for 24 hours.

Dimensions

The visual inspection includes girth tape measurements along the pipe. Normally, a maximum outer diameter and a minimum submerged weight will have been specified (for on-bottom stability), and the coater will have determined a nominal concrete thickness based upon the design concrete density. The concrete coating should be concentric with the pipe, and free of excessive undulations, typical tolerances on the nominal coated pipe diameter being −10 mm to +20 mm.

Interface friction

If the pipeline is to be installed by conventional pipelaying, it is important that the concrete coating cannot slip over the anti-corrosion coating, which is documented by push-off tests on the finished coating. A length (typically 1.5–2 m) of concrete coating at each end of a pipe joint is separated by a circumferential saw cut, and pushed off by hydraulic jacks, the required failure strength depending upon the envisaged laybarge tensioner force; see Figure 7.23.

Figure 7.23   Push-off test on concrete coated pipe

Push-off tests are normally carried out at room temperature, but for asphalt enamel coatings to be used at the upper limit of the temperature range (see Section 7.4.3) it might be appropriate to verify the shear strength at the topical temperature. Cases have been recorded of steel pipes creeping out of concrete coatings restrained by soil friction.

The interface shear strength between concrete and hot applied enamel coatings will normally be sufficiently large to ensure that failure occurs in the anti-corrosion coating, which has a shear strength of approximately 0.1 MPa at ambient temperature.

For coatings such as FBE or three-layer polyolefins, which are more slippery, special measures must be introduced, such as mechanical roughening, embedding of sand grains, etc.

Impact resistance

The ability of the concrete to resist impact from fishing gear should be documented by impact tests. A pipe joint is adequately supported, e.g. by a sand berm or a massive rig, and hit with a hammer at a specified impact energy; see Figure 7.24. The mass and striker edge of the hammer represent typical trawl equipment, and common tests are:

Figure 7.24   Rig for impact testing of concrete coating

Figure 7.25   Narrow striker edge of 2680 kg impact hammer

- 75 mm flat face hammer of 1000 kg, travelling at 2.0 m/s (4 knots);
- 10 mm radius hammer of 2680 kg, travelling at 2.76 m/s.

For the former test the acceptance criterion would be that no reinforcement is exposed by sixty repeated blows at the same spot. For the latter more severe test the anti-corrosion coating should suffer no damage, and the radius of spalling should not exceed 300 mm, after five blows. The impact angle is at 90° to the pipe axis, but oblique impact testing (e.g. at 60°) may also be specified. The striker edge of the above heavy hammer is shown in Figure 7.25, and Figure 7.26 shows a concrete coating after having received five blows.

7.7.9   *Anode joints*

Anodes are normally placed on the pipe joint prior to the application of a concrete coating, and shielded from contamination by cement or concrete. For concreting by slipforming it may be more practical to install the anodes after-wards. In any case, if the thickness of the concrete is greater than that of the

Figure 7.26   Concrete coating after five blows from a heavy impact hammer (no anti-corrosion coating exposed)

anodes (which is typically 40 mm), the concrete coating is tapered down to the anode; see Figure 7.15.

7.8   Marking, handling and repair

7.8.1   *General*

The finished, coated pipe joints are stored at the coating yard, or possibly an intermediate storage area, before being loaded out to the pipeline installation contractor. Procedures for marking, handling and repair are needed to ensure that the pipe remains identifiable and fit for installation.

7.8.2   *Marking*

The purpose of the marking is to allow unambiguous identification of the individual pipe joint, including its history of coating, repair, and anode attachment. At the pipe mill each joint is assigned a unique pipe number, which is die stamped at the bevel faces and stencil painted on the inside bore. The paint

Figure 7.27 Internally coated pipe with identification marking

stencil also includes information of the pipe outer diameter (OD), wall thickness (WT), and the length of the joint (in cm).

After internal coating (if suitable) the number and length is transferred (by felt pen or similar) to the coating; see Figure 7.27. It is important that this information is provided in time and at both ends, to be accessible after possible double-jointing at the laybarge, as the die stamps may soon become illegible during outdoor storage. All information on the pipe coating (the type of anti-corrosion coating, attached anode half-shell numbers, thickness of concrete coating, weight of the joint, repair details, etc.) is recorded (in electronic and paper files) against the pipe number, for later retrieval. Similarly, the sequential pipe number assigned to the joint at installation is painted on the outside of the pipe.

7.8.3 *Handling, transport and storage*

To avoid damage to the pipe steel or coating, pipe joints should not be dragged, but picked up with wide, non-abrasive belts or brass-lined end hooks. In the latter case a spreader bar should be used between lifting lines. Some operators specify the use of bevel protectors, but they offer little protection against pipe end damage, tend to retain moisture, and are a nuisance at the coating yard or at the construction site.

Bare or coated pipe transported on flat-bed trucks or rail cars should be adequately secured and padded, and pipe on vessels must be secured against rolling.

Pipe should be stacked in such a way as to prevent damage to the pipe or its coating, and the safe stacking height be documented by calculations. Normally concrete coated pipe is stacked up to nine layers high, with the bottom row on sand berms to avoid contamination from the ground, see Figure 7.2. Anode joints are often stacked separately, or in the top row and, in any case, such that no weight is bearing on the anode.

Prolonged outdoor storage of coated pipe should be avoided, as it has been known to compromise the integrity of internal flow coatings, and lead to excessive rust formation if there is no internal coating. The closure of the pipe interior by means of end caps is not recommended, as the moisture tightness of the caps cannot be guaranteed.

7.8.4 *Repair*

Damage to pipe ends, such as dents or gouges, require the removal of pipe material. When a section of pipe is cut off, the new end should be ultrasonically inspected along a certain distance (typically 100 mm) to detect any lamination or defects. If the cut-off length exceeds 150 mm it may also be necessary to remove some coating. The pipe identification must be transferred, and the new pipe joint length recorded.

Repair criteria and procedures for anti-corrosion coating depend upon the type of coating; see Section 7.4 above.

Minor damage to the concrete coating can be tolerated, and normally a loss of not more than 25% of the concrete coating thickness in an area with a diameter not exceeding 15% of the pipe diameter will be acceptable. In the case of greater damage the coating should be chiselled out to provide a key lock to sound concrete. If the area is ten times larger it is recommended that the coating around the entire perimeter of the pipe is removed. Repairs are normally carried out using the same concrete as the original coating (though possibly a somewhat wetter mix), but special, rapidly setting, cementitious repair materials may also be used.

Annual cracks at the pipe end that are over 30 mm deep should be repaired, and the same goes for larger cracks that may arise from pipe handling. However, circumferential cracks that are less than 1.5 mm wide and longitudinal cracks that are less than 300 mm long should be tolerated. Unacceptable cracks should be chiselled out and repaired.

7.9 Components and fittings

7.9.1 *General*

Components comprise all the appurtenances that are connected to the pipeline, either concurrently with the installation or prior to commissioning. Single items (hubs, flanges, tees, wyes, valves, isolation couplings, etc.) are normally steel forgings, and larger assemblies (swan necks, spools, pig launchers/receivers, slug catchers, etc.) are welded together in onshore shops.

All items are protected against corrosion by a supplier specified coating, typically an epoxy paint system, and the welds are normally protected by heat shrink sleeves. Alternatively, the individual components may be delivered bare, and the completed assembly coated by the manufacturer. Spools will often include regular pipeline joints with the corresponding anti-corrosion (and possibly concrete) coating.

Single bends welded into the pipeline, for example as part of an expansion offset, will typically be hot formed (see Section 7.2.6), and are not considered components. Forged bends and elbows are normally welded into larger components or assemblies. All components are provided with pup pieces to facilitate welding to adjoining linepipe. Pup pieces may be welded on by the component manufacturer, or they may be integral with the forged item.

7.9.2   *Component manufacture*

Flanges are normally off-the-shelf items, ordered to standard specifications, depending on the pressure rating and the transported medium. Hubs, tees and wye pieces are manufactured by specialty suppliers. All items (except blind flanges) are provided with weld necks or pup pieces for welding the component into the pipeline or assembly.

Valves are manufactured by specialty suppliers to specifications describing the pressure rating and mode of operation, and provided with weld necks for connection to spool pieces or pup pieces. In addition to the main line valves, smaller bore valves may be needed for by-pass lines, venting, etc.

An isolation coupling may be a pair of flanges provided with a special isolating gasket, or a prefabricated monoblock isolation joint, manufactured by a specialty supplier, ready for welding into the pipeline or assembly. For horizontally placed isolation couplings in particular, there is a risk of short-circuiting by deposits or debris in the pipeline, and monoblock joints tend to be more reliable.

7.9.3   *Risers and expansion offsets*

The riser is the vertical (or almost vertical) part that connects the pipeline on the seabed with the topside equipment of an offshore platform. At the foot of the riser an expansion offset may be introduced to absorb the movements of the pipeline end due to changes in temperature and pressure. Depending on the orientation of the pipeline approach to the platform, the expansion offset may be an L-, Z- or U-shaped spool; see Figure 7.28.

To avoid overstressing the linepipe steel, it may well be a requirement that the spool remains above the seabed to reduce the passive soil resistance to temperature movements; see Section 6.5.2. To prevent embedment, the offset may be provided with wear rings, sliding on skid beams placed under the spool; see Figure 7.28.

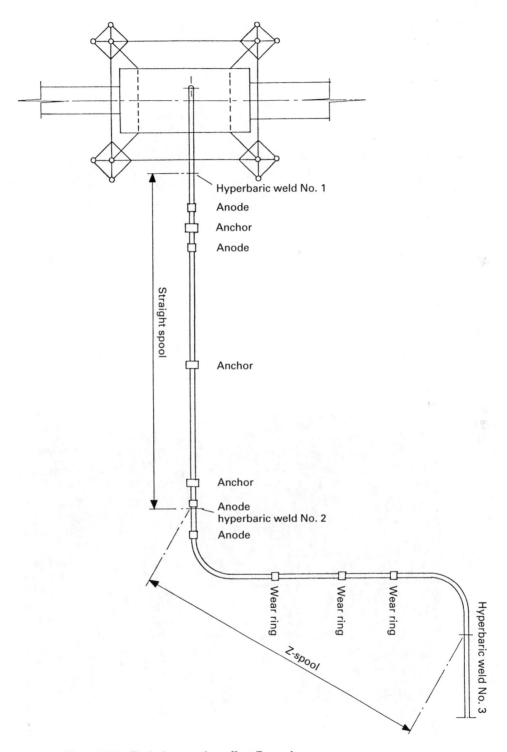

Figure 7.28   Typical expansion offset Z-spool

Risers and expansion spools are welded up from straight linepipe joints and hot formed (or forged) bends, provided with anti-corrosion coating. To avoid damage to the riser in the splash zone and the exposed expansion spool on the seabed the preferred choice of anti-corrosion coating is Neoprene (polychloroprene) elastomer. The girth welds will normally be protected by heat shrink sleeves, whether they are applied in the shop or as offshore field joints; see Section 8.3.3.

7.9.4 *Tees and valve assemblies*

A tee assembly consists of a tee piece, welded into the pipeline, and a short branch line, equipped with an isolation valve and/or a blind flange.

Figure 7.29 shows a typical tee assembly, incorporating (from right to left):

- forged tee piece;
- forged 1.5D elbow;
- ball valve;
- forged tee support piece, with tee support cradling the branch line;
- tie-in porch and hub for tie-in of the branch line; see Section 8.6.4.

A valve assembly may be tied in to the pipeline directly, or placed on a riser base, see Figure 11.1. As described in Section 6.8.3, the assembly comprises one

Figure 7.29 Typical tee assembly

or two valves, plus the necessary by-pass lines. A small bore by-pass may be built into the valve body, otherwise a by-pass line, provided with the appropriate small bore valves, must be installed. The connection to the main pipeline is typically reinforced by an olet, which is a collar-shaped fitting. The hole, which may be torch-cut or drilled, can be made before or after the fitting is welded on to the pipeline. The latter solution, which is suitable for sizes not less than 2″, prevents any distortion of the main pipe. The branch line is normally welded to the fitting (weldolet), but alternatively (for low pressure lines) the fitting may be provided with threading (thredolet). Usually the olet is normal to a straight pipeline, but it can also be oblique (latrolet) or placed on a bend (elbolet).

Weldolets are typically specified whenever small bore access to the pressurised pipeline is required, e.g. for pig signallers, vent or drain panels, etc. However, no matter how carefully they are manufactured, the connections are vulnerable to impact during installation and vibrations during operation, and are as far as possible avoided by built-in bypasses, non-mechanical pig-signalling, etc.

7.9.5 *Pig launchers and receivers*

Normally facilities for the launching or receiving of pigs will be located onshore or at platform topsides, and thus not form part of the offshore pipeline. However, in connection with subsea completions or with spur lines, such items may be installed as pipeline components on the seabed. The complete assembly, including valves, by-bass lines, pig signallers, etc. is welded up onshore, provided with corrosion protection, and installed like any other valve assembly.

Chapter 8
Installation

8.1 Introduction

Marine pipeline installation comprises all the activities following the fabrication of the pipe joints, bends and components through the preparation of the pipeline for commissioning. The principal exercise is the joining of the individual pipe joints into a continuous pipe string. This may take place concurrently with the installation on the seabed by laybarge, or it may be carried out onshore in preparation for installation by reeling, towing, pulling or directional drilling. To construct the complete pipeline it may be necessary to perform offshore tie-ins to other pipe strings or to risers. These connections may be carried out on the seabed or above water.

The basis for the installation activities is the route survey, as documented on the alignment sheets, which also specify the principal features of the pipeline; see Section 2.6.2. When the pipeline is installed the actual length is recorded by accumulating the lengths of the installed pipe joints. Owing to installation deviations and seabed irregularities the distance (chainage) measured along the pipeline will be slightly higher than the kilometre post (KP) designation measured along the theoretical route, and given on the alignment sheets. Thus if KP numbers are used to indicate chainage it should be highlighted to avoid confusion.

To avoid overstressing or excessive displacements, seabed intervention may be needed prior to or concurrent with the placement of the pipe string. Subsequently, post-trenching and/or backfilling may be required to ensure long-term hydrodynamic stability, upheaval buckling resistance, or mechanical protection.

Once the pipeline is in place on the seabed, the integrity is documented by internal gauging and hydrostatic testing, which requires filling of the pipeline with water. Subsequently the pipeline is made ready for service by removal of the test water, which in the case of gas pipelines includes complete drying.

8.2   Seabed intervention

8.2.1   *General*

Modifications of the natural seabed that may be needed include:

- soil replacement to improve the foundation properties, for example, to prevent the pipeline from sinking into soft mud;
- trenching to reduce the actions from waves and current;
- protection of existing pipelines or cables in connection with crossings;
- provision of supports for riser bases, spools, etc.;
- reduction of free span heights to reduce the forces due to overtrawling;
- smoothing of the pipeline profile to reduce the length of free spans or prevent contact pressures that could damage the coating or dent the pipe steel.

Spans that are unacceptable in the unstressed, airfilled condition must be rectified before installation of the pipeline (pre-lay intervention). Other free span rectification may be postponed until after the pipe string has been placed (post-lay intervention).

For fatigue verification it is common industry practice to require that no more than 10% of the allowable damage ratio (see Section 6.4.6) be reached during the temporary installation phases, which include a period in which the pipeline is empty on the seabed and a period in which it is waterfilled. As the free spans are typically larger in the former case, the accumulated damage can be reduced by flooding the pipeline shortly after installation, thus obviating the need for pre-lay span rectification.

Once the pipeline is ready for operation it will be known what fatigue damage may have occurred during installation (pipelaying) and for how long the free spans will have been exposed in the empty and in the waterfilled conditions (it is not necessarily the same pipeline sections in the three cases). The remaining fatigue life of the free spans in the operating condition can then be calculated, and post-lay free span rectification carried out if necessary.

Trenching and backfilling after pipe string installation are treated in Section 8.7, whereas pre-lay seabed intervention is discussed below.

8.2.2   *Pre-trenching*

The construction of pipeline trenches or the removal of outcrops to reduce free spans are normally achieved by dredging. Depending upon the type and hardness of the seabed soil different dredger types (bucket dredger, cutter-suction dredger, etc.) are deployed. In shallow waters the dredging may be carried out from jack-up platforms, or even fixed structures. In extreme cases blasting with explosives may be necessary.

As the trench must be sufficiently wide to receive the pipeline, pre-trenching is an attractive option only for pipelines installed by towing or pulling, including shore approaches, or for laying in very shallow water or across mud plains, where trenching is needed to gain access for the pipelay vessel. Such near-coast areas are also the most environmentally sensitive, and pre-trenching is likely to be subject to severe restrictions, including those for the disposal of the dredged material.

8.2.3 *Pipeline supports*

As an alternative to the removal of seabed material a suitable support for the pipeline may be created by rock dumping, either along the entire section affected, or – more likely – as isolated gravel berms.

The size of the individual stone or gravel particles shall be adjusted to the environmental conditions to ensure that the material will not be removed by wave and current action. The rock dumping may be performed by split barges, but much more economical use of material is obtained by using a rock dumping vessel equipped with a fall pipe, through which the material may be placed over the pipeline with great accuracy. Fall pipe dumping is routinely performed at water depths exceeding 300 m, the vertical tolerance being ± 200 mm.

Pre-lay free span supports will have to be made sufficiently wide to cater for the pipelay tolerance as well as the horizontal tolerance on rock dumping, which is greater when the gravel berm cannot be related to a fixed object (e.g. the pipeline) on the seabed. Thus there is an economic incentive to use post-lay rather than pre-lay intervention; see Section 8.2.1 above.

Unacceptable free spans may also be reduced by structural supports, established on the seabed during the installation. Proprietary systems have been developed where the pipeline can be supported at variable height above the seabed, and the supports can be placed by divers or diverless vehicles etc. This may be the only alternative for very large water depths, where neither dredging nor rock dumping is viable. Isolated supports made from grout bags or similar may also be installed under the pipeline to prevent it from sinking deeply into very soft seabed soil such as organic mud.

It may also be necessary to establish lateral supports to guide the pipeline in the horizontal plane to achieve the desired lay radius if the lateral soil friction is insufficient; see Section 6.3.6. Rather than gravel berms, such counteracts may take the form of structural elements that can be retrieved and reused. A typical design would be a hollow concrete cylinder, possibly provided with a steel skirt that penetrates the seabed to enhance the lateral resistance. To increase the support capacity further the cylinder can be filled with gravel.

8.2.4 *Crossings*

The pipeline may well have to cross existing pipelines or cables identified by the route survey. If these are abandoned the only concern is that they may prevent trenching of the pipeline. The crossing of live pipelines and cables, on the other hand, calls for special measures, including negotiations with the third party owners. To avoid damage to any of the installations the lines should be separated by a suitable material, and often the relevant authorities will have requirements for the crossing design. DNV OS-F101 specifies a minimum separation of 0.3 m.

The existing cable/pipeline may already be trenched into the seabed, or it may be allowed and feasible to carry out such trenching. In that case is suffices to lay the new pipeline on top. As it cannot then be trenched, it may have to be protected by post-lay rock dumping or similar.

In most cases, however, the existing cable or pipeline will be lying on the seabed and cannot be touched. Then it is necessary to engineer a pipeline crossing using rock berms, mattresses, grout bags or similar, so the new pipeline can be laid across without damage to the existing one, and the entire crossing covered, if required.

8.3 Pipe assembly

8.3.1 *General*

The individual pipe joints are assembled into pipe strings by girth welding, i.e. circumferential butt welding. Strict requirements are specified for welding consumables, welding procedures, welder qualifications, as well as inspection and testing, in order to ensure optimum quality of the pipeline.

The welding can be carried out offshore during pipelaying (on the laybarge or pipelay vessel), often referred to as marine welding. Pipe strings may also be made onshore for subsequent offshore installation by reeling, towing, pulling and directional drilling. This is then referred to as site or shop welding.

Onshore fabrication of the pipe string includes the field joint protection, and for towed or pulled pipelines it also includes the attachment of sacrificial anodes. It is also possible to insert a thermoplastic liner for internal corrosion prevention and, basically, two methods are used. A liner with an outer diameter slightly greater than the pipeline bore may be pulled into the string through a diameter reduction device, and when the tension is removed the liner relaxes back to a close fit. Alternatively, a liner with a diameter slightly less than the pipeline bore may be pulled in, and subsequently expanded by pressurised hot water. Owing to the lower friction, the latter method can be used for pipe strings up to 1000 m in length. However, as mentioned in Section 6.6.3, for hydrocarbon pipelines special measures must be taken to prevent diffused gases from collapsing the liner in the case of depressurisation. Consequently, liners are mostly used for water lines. The corrosion protection properties may be enhanced by injecting cement paste between the liner and the steel pipe. An even thickness of the cement layer is ensured by the Velcro type external surface of the collapsible liner, which is pressed against the pipe wall by pulling a soft, but snugly fitting plug through the pipe string.

The installation of a long pipeline often involves the assembly of several pipe strings. On a reel barge a new reel is joined to the pipeline in much the same way as for laybarge installation, usually after recovery of the previous string from the seabed. Assembly of two pipe strings left on the seabed may be accomplished by either mechanical coupling devices or by welding; see Section 8.6.4. The welding operation may be carried out above surface (davit lift), or subsea (hyperbaric welding). In the latter method the welding is carried out in a pressurised habitat placed over the ends of the pipeline that are to be joined. Mechanical coupling devices may be substituted for welding in such cases; for deep water this is the only option.

8.3.2 *Girth welding*

Joining the pipe by girth welding consists of circumferential welding. The abutting pipe ends are prepared to facilitate easy welding according to the welding process to be used. The processes considered for modern girth welding of pipelines are:

- friction welding;
- explosion welding;
- electron beam welding;
- laser welding;
- submerged arc welding, SAW;
- shielded metal arc welding, SMAW (using cellulosic or low hydrogen basic electrodes);
- gas metal arc welding, GMAW.

The first four methods have not yet made it beyond the novelty stage, whereas the latter three are used extensively.

Pipeline girth welding is covered by a number of standards, such as:

- API 1104;
- BS 4515;
- BS 7910;
- CSA Z184.

These standards are devoted to all aspects of welding, including non-destructive testing and acceptance criteria for such testing. The acceptance criteria are basically written down as 'Good Workmanship Standards', such as API 1104, and the defect acceptance is essentially derived from radiographic images of the weld, i.e. length, width and type of defect. The workmanship criteria can be 'transformed' into ultrasonic inspection by assuming that the acceptable height of the defect corresponds to the typical height of a single pass in multipass welding, i.e. 3 mm. This philosophy may be sufficient for manual welding, but as it is being extrapolated to mechanised welding the acceptance criteria will be better based on a 'Fitness for Purpose' concept, or rather on so-called engineering criticality assessment (ECA). This is already incorporated in the above-mentioned standards under certain conditions, but more recently the EN 1594 and the DNV OS-F101 deal in great detail with ECA based acceptance criteria, as does the European Pipeline Research Group: *Guidelines for defect acceptance levels in transmission pipeline girth welds*. In Section 8.3.3 below an example is given of an ECA based upon BS 7910.

To increase the efficiency in pipe laying it is customary to pre-make double joints in a separate welding operation, either onshore or – most likely – on the laybarge, and then to add these double joints to the pipeline in a semi-continuous process in the so-called 'firing line'. The double joint welding is usually fully mechanised. There is normally sufficient time to conduct the welding and to perform inspection, testing and possible repair welding before entering into the final production line. The preferred welding process depends on the linepipe material. For carbon steel linepipe materials the standard choice is SAW, as it is a reasonably fast and reliable welding process, but it may be combined with root and hot passes using mechanised GMAW. For corrosion resistant alloys the welding process will be some form of semi- or fully mechanised GMAW. Carbon steel with internal lining or cladding of corrosion resistant alloys may use internal or externally applied GMAW in the root, with SAW to fill up the weld gap.

In the firing line the normal choice today will be mechanised GMAW, but SMAW is also used. Until 1970, marine girth welding was dominated by SMAW using cellulosic electrodes to weld vertically down at a high deposition rate. Their main drawback is a high propensity to hydrogen cracking when the welds cool down, but careful control of welding procedures enables control of the cracking at least up to a strength grade suitable to weld X65 pipes. For higher grade steels, or if overmatching welds are required, the next step up is the use of basic covered electrodes and, recently, also fuse cored electrodes. This enables high strength weld material to be used without the risk of hydrogen cracking. The demand for higher productivity led to the development of mechanised GMAW. Not only is it possible to cut down the welding time considerably, but it also removes the risk of hydrogen cracking and allows a broad spectrum of high strength weld metals to be used. Mechanised welding generates only a few random defects, but lends itself to producing repetitive and systematic flaws if it is not continuously monitored and checked. Feed-back to the welder needs to be fast in order to avoid extensive repairs, thus mechanised welding has also provided a challenge to the non-destructive testing of welds. Today mechanised welding goes hand in hand with automated ultrasonic inspection and special acceptance criteria to give a complete or holistic approach to girth welding.

The firing line weld is made with a root pass and one or more hot passes before the filler passes are deposited. There are several proprietary mechanised welding processes available, using more or less narrow gap welding techniques to cut down

Figure 8.1   Field joint welding on laybarge (photo: DONG A/S)

the amount of deposited weld metal (and hence welding time). Some of the processes make use of an internally deposited root pass, while others are completely single-side welded. Typical semi-automatic field welding is shown in Figure 8.1.

The firing line welds are inspected and tested before leaving the barge. If non-acceptable defects are detected, the line is pulled back and the weld is repaired by either a complete cut out or an approved local weld repair procedure.

It is very important to consider in advance not only the type of inspection and testing to use to ensure the weld integrity, but also what the acceptance criteria will be for the selected testing method.

Visual inspection is still used widely today as it offers a rapid check for arc burns/strikes, weld crown irregularities, undercuts, bead concavity, etc. If possible it should also be carried out from the internal side.

X-ray inspection detects volumetric defects such as porosities, and will also pick up inclusions, burn through, root bead concavity and lack of penetration. If planar defects are lined up in the direction of the x-rays they will also be detected.

Manual ultrasonic testing using conventional pulsed echo probes can pick up most of the defects, except porosities and root bead concavity. The method is very time consuming because it is necessary to manipulate the probes extensively to cover the entire weld volume, and it is necessary also to use several probes with different angles to cover all types of defects.

To cut down the inspection time the ultrasonic testing has been mechanised. Several probes are used simultaneously with different angles and different depths of focus. In this way it is possible to tailor-make the inspection to specific weld bevel geometries. By also using phased array probes and time of flight diffraction techniques it is possible to cover the entire weld zone and all type of defects with a high probability of detection in one go around the weld, bringing down the inspection time to a few minutes, and with instant feedback of the examination result to the welder. These highly sophisticated inspection processes are dependent on calibration and validation prior to on-line inspection, and also on frequent calibration during the inspection work.

The industry has over many years worked to the 'Good Workmanship Standard' using visual inspection and radiography (x-ray). The workmanship criteria are based on the weld quality that can be expected from a good welder (on a good day) and working to these criteria the repair rate is generally below 5%. The experience gained (or rather the lack of girth weld failures experienced) justifies that pipelines exposed to (mainly) static stresses below the specified minimum yield of the pipe material are checked against such 'workmanship' criteria.

For pipelines exposed to plastic strain during pipe lay or during subsequent installation processes, such as trenching/ploughing or davit lifting above the sea surface, or pipelines exposed to excessive free spanning with high bending strains and possible fatigue due to vortex shedding, etc., the workmanship criteria may not be sufficient. In such cases the acceptance criteria should be derived from fracture mechanics based engineering critical assessments, taking into consideration such factors as:

- the mechanical properties across the welds (undermatching or overmatching weld strength);
- the local geometry of the weld;
- the amount of strain;
- the strain rate;
- the stress/strain amplitude;
- the possible influence from the external and internal environments in terms of either active corrosion or hydrogen uptake.

Based on the outcome of the ECA it is then possible to define position, length and depth of acceptable defects, and to select the non-destructive test method(s) that will find non-acceptable, harmful defects with a satisfactory probability of detection.

The DNV OS-F101 provides comprehensive coverage of pipeline welding and non-destructive testing requirements.

Hyperbaric welding, inspection and testing should follow the same principles as above. Welding procedure testing should be made on pipe joints under conditions duplicating the actual physical conditions in the habitat and in the pipes. The hyperbaric welds should be made with the same equipment and by the same welders/weld operators as the test welds. The weld inspection and non-destructive testing should be made in a such a way and by such methods that will result in the same or better probability of detection of harmful defects. Weld repairs are only allowed if the weld repair procedures are qualified and validated through the same tests as for the girth weld.

8.3.3   *Engineering criticality assessment (ECA)*

Most codes and standards request that acceptance criteria for non-destructive testing (NDT) of pipeline girth welds are based on an engineering criticality assessment (ECA) when the accumulated plastic strain resulting from installation and operation is above 0.3%. The process of developing such acceptance criteria is illustrated in Figure 8.2.

Basically it requires an ECA and qualification of the NDT equipment and procedures in order to develop the acceptance criteria. The engineering critical assessment provides a set of curves describing the limits for critical defect sizes, and the qualification gives information about the detection and sizing capabilities of the NDT system. The ECA may be performed using BS 7910 to determine critical defect sizes. If there are in-service conditions inducing failure by fatigue or unstable crack growth the assessment shall be at Level 3, otherwise a normal assessment, Level 2A, is sufficient.

The ECA is based on actual material specifications and actually measured fracture toughness properties (crack tip opening displacement (CTOD), critical stress intensity factor, K_{IC}, or J-integral) of the weld metal, heat-affected zone

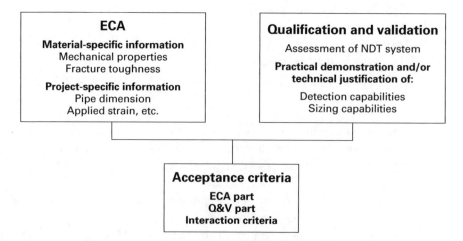

Figure 8.2   Schematic showing the development of acceptance criteria for NDT

and base metal from the welding procedure qualification. Due consideration should be given to the possibility that some pipes may have lower fracture toughness than the pipe used in welding procedure qualification testing. Also, repeated repair welding must duly be considered as it may impede the fracture toughness considerably, as may high heat input welding methods used for double jointing outside the firing line.

An example of ECA is shown below in order to illustrate the influence of fracture toughness values on the critical defect sizes. The data for applied strain and CTOD-value are maximum and minimum values, respectively, therefore no safety factors have been used in the calculations. The maximum value for applied strain is assumed to include all stress and strain concentration factors.

Design basis and assumptions: preliminary ECA

Pipe inner diameter, D_i	729 mm
Nominal wall thickness, t	17.5 mm
Material	API 5L X65
Young modulus, E	207 000 MPa
Poisson coefficient	0.3
SMYS	450 Mpa
SMTS	520 MPa
Strain at ultimate stress	23%
Minimum weld material fracture toughness, (here crack tip opening displacement)	0.2–0.5 mm
Maximum applied strain	0.5%
Embedded defects	2 mm from surface

Notation

a	defect depth
c	defect half length
d	equal to a for surface defects, and $2a$ for embedded defects
K	stress intensity factor
K_{mat}	material toughness measured by stress intensity factor
K_r	fracture ratio of applied elastic K value to K_{mat}
$M_{m,b}$	stress intensity factor, membrane/bending
$M_{km,b}$	stress magnification factor, membrane/bending
P_b	primary bending stress
P_m	primary membrane stress
Q_b	secondary bending stress
Q_m	secondary membrane stress
R_m	middle radius for pipe
S_r	load ratio of applied load to flow strength load
L_r	load ratio of applied load to yield strength
Y	stress intensity factor correction

Φ mathematical factor (2nd order elliptical integral)

β c/R_m

δ crack tip opening displacement (CTOD)

δ_I applied CTOD

δ_{mat} material toughness measured by CTOD method

δ_r fracture ratio using CTOD parameters

ε_u strain at fracture

ε_y strain at yield

η $1 - d/t$

ρ plasticity correction factor

σ applied stress

σ_{nom} applied nominal stress

σ_f flow strength

σ_K effective net section stress

σ_u tensile strength

σ_Y lower yield strength or 0.2% proof strength

Methodology applied to calculate critical defect size
The steps in the calculation method are as follows.

The assessment curve is given by:

$$\sqrt{\delta_r} = K_r = (1 - 0.14\,L_r^2)[0.3 + 0.7\,\exp(-0.65\,L_r^6)]$$

where $L_{r,max} = 1.15$ (taken from BS 7910:1999 Figure 11a)

$$\sqrt{\delta_r} = \sqrt{\frac{\delta_1}{\delta_{mat}}} + \rho$$

$$\delta_1 = \frac{(Y\sigma)^2\,\pi a}{\sigma_y E}$$

ρ is given in BS 7910 annex R.

$$Y\sigma = (Y\sigma)_p + (Y\sigma)_s$$

$$(Y\sigma)_p = \frac{1}{\Phi}\,(M_{km}M_m P_m + M_{kb}M_b P_b)$$

$$(Y\sigma)_s = \frac{1}{\Phi}(M_m Q_m + M_b Q_b)$$

M_m and M_b are determined according to BS 7910 Annex M3.
M_{km} and M_{kb} are determined according to BS 7910 Annex M5.

Φ is a correction factor defined in BS 7910 Annex M, formula M9

$$a/2c < 0.5; \; \Phi = [(a/c)^{1.65} \, (1 + 1.464)]^{0.5}$$

$$0.5 < a/2c < 1; \; \Phi = [(c/a)^{1.65} \, (1 + 1.464)]^{0.5}$$

The effective net section stress σ_K for a surface circumferential defect under pressure is determined according to Kastner *et al.* (1981) (see also Miller 1988, p. 291):

$$\sigma_K = \sigma_{nom} \Bigg/ \left\{ \frac{\eta[\pi - \beta(1 - \eta)]}{\pi\eta + 2(1 - \eta)\sin\beta} \right\}$$

The plastic collapse ratio is determined by:

$$L_r = \sigma_K / \sigma_y$$

Primary stress

The primary axial load of the pipe is given by the limit strain level, ε_m. The corresponding stress level, σ_m, is found using the Ramberg-Osgood curve for the relevant material:

$$\varepsilon_m = \frac{\sigma_m \cdot \left[1 + \alpha \left(\dfrac{\sigma_m}{\sigma_y} \right)^{(n-1)} \right]}{E}$$

This stress level is used as the primary membrane stress P_m.

The values for α and n for the material are determined by the formulae:

$$\alpha = \frac{\varepsilon_y \cdot E}{\sigma_y} - 1$$

$$n = \ln\left[\frac{(\varepsilon_u \cdot E) - \sigma_u}{(\varepsilon_y \cdot E) - \sigma_y} \right] \ln\left(\frac{\sigma_y}{\sigma_u} \right)$$

Secondary stress

The residual stress due to the welding is assumed to be the lesser of:

$$\sigma_y \quad \text{and} \quad \left(1.4 - \frac{\sigma_k}{\sigma_f} \right) \sigma_y \quad \text{(from BS 7910 7.3.4).}$$

The secondary stress, Q_m, is set equal to the calculated residual stress.

Table 8.1 Calculated critical defect heights as function of defect lengths and CTOD-value

Defect type	Defect length (mm)	Defect height (mm)			
		δ = 0.2	δ = 0.3	δ = 0.4	δ = 0.5
Surface breaking	20	2.2	4.9		
or interacting	30	2	3.7	5	6.6
	40	1.9	3	4.4	5.4
	50	1.83	2.9	4.1	4.9
	70	1.78	2.7	3.8	4.4
	100	1.7	2.6	3.3	4.1
	150	1.64	2.4	3	3.8
	200	1.6	2.3	2.9	3.4
Embedded	20	5.15	8	11.6	13.9
	30	4.55	6.3	8.1	9.8
	40	4.3	5.8	7.1	8.3
	50	4.2	5.5	6.65	7.7
	70	4	5.2	6.1	7
	100	3.8	4.9	5.7	6.4
	150	3.6	4.5	5.3	5.9
	200	3.45	4.3	4.9	5.5

Calculated critical defect heights for given defect lengths
Table 8.1 and Figures 8.3 and 8.4 illustrate the defect length, defect height and the corresponding fracture type for different defect types calculated for different CTOD-values.

The curves illustrate that the better the material is in terms of fracture toughness, the larger the defect that can be tolerated in the welds. In the ideal case the

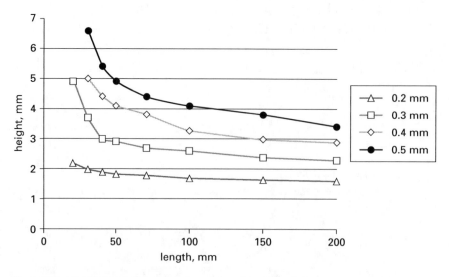

Figure 8.3 Critical defect heights for surface breaking defects

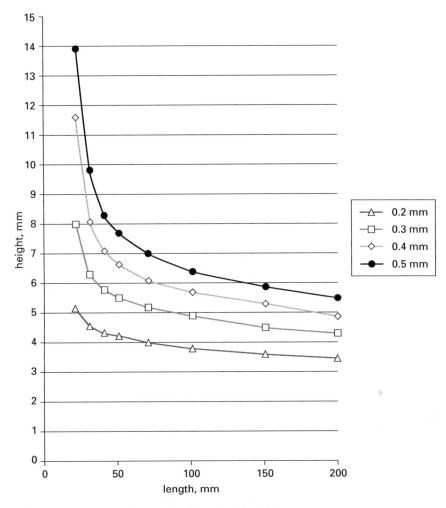

Figure 8.4   Critical defect heights for embedded defects

acceptable defect size is the same as the critical defect size. In reality acceptable defects should be chosen to be smaller than critical ones to cover for uncertainties due to the following.

Defect sizing capabilities

A state of the art NDT technique allows for very accurate height sizing of defects. The tolerance on the height sizing capability of the applied NDT system is verified in the qualification of the equipment, and can therefore be accounted for when establishing a set of allowable defect sizes.

Probability of detection aspects

A semi-automated welding technique combined with a state of the art NDT system (e.g. automated ultrasonic examination system), that is tailored to detect the expected defects, ensures a very high probability of detection of the critical defects.

In addition to this, the defect population distribution is not known at the time when the acceptance criteria are developed, thus the probability of occurrence of defects cannot be quantified up front. This problem is addressed by imposing a quality assurance that focuses on the accumulated defect length within welds and within a number of consecutive welded butt joints. These limits ensure that the quality of the welding remains high, and they have, in principle, nothing to do with the integrity of the weld.

8.3.4 *Field joint coating*

After welding the linepipe steel, the weld area and the adjacent coating cut-back is protected by field joint coating. Corrosion protection is normally achieved with tape wrapping or heat shrink sleeves; if both the adjacent pipe joints are provided with concrete coating or insulating coating a suitable infill material is applied to the field joint area, to create a continuous pipe string. A typical field joint on a concrete coated pipe is shown in Figure 8.5.

Welds to pipe joints or bends that are not concrete coated, as well as to components (weld neck flanges, valves, isolation couplings, etc.), are normally provided with heat shrink sleeves, or if produced in a shop they may be coated with an epoxy paint system, similar to that used on the corresponding component; see Section 7.9.1.

If the adjacent pipe joints or components are provided with fusion bonded epoxy coating (FBE, see Section 7.4.5) the same coating may be applied to the

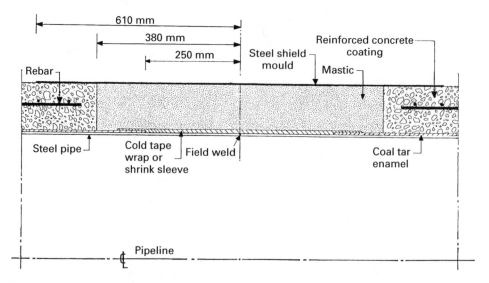

Figure 8.5 Typical pipeline field joint

field joint. This involves grit blasting to a uniform grey appearance (surface quality SA 2.5 according to ISO 8501-1) and heating, normally by induction, of the pipe wall to the temperature (typically 220°C) needed for gelling and curing of the epoxy powder, which is sprayed on the joint area.

If FBE is used as an internal anti-corrosion coating, a similar procedure must be applied to the internal surface of the girth weld. Proprietary systems have been developed for this purpose, and although their use is too time-consuming for offshore application they can be used for the onshore manufacture of pipe strings.

Field joints in three-layer polyolefin coatings may be performed to the same standard as the parent coating. The application of FBE primer (typically 150 micron) is then followed by spray application of adhesive and the injection of polyolefin into a mould enclosing the girth weld area. The resulting field joint coating will be somewhat thicker than the adjacent pipe coating, and reported application times are 6–7 minutes. The factory-applied coating should be chamfered to a low angle (20°–30°) to reduce the shear and normal stresses on the interface when the pipeline is bent during laying or reeling. In particular, the normal stresses would tend to lift the coating off at the edge, where the bond may be weakened due to the heat from the girth welding. Similar injection moulding techniques are used for coating field joints in pipelines insulated by syntactic foam; see Section 7.4.7.

For the application of tape wrap or heat shrink sleeves the surface preparation is limited to cleaning and wire brushing to surface quality SA 3, and the total installation time is 3–4 minutes.

Wrapping tape may be fabricated from PVC or polyethylene, typically 0.75 mm thick, with a self-adhesive layer, normally a butyl rubber compound. The total thickness is 1.5–2.0 mm. The tape is cut to size, and applied to the field joint as a cigarette wrap, with the overlap at the top of the pipe. It is customary to use three wraps, one to cover the exposed steel and two (narrow) wraps to bridge the gaps to the adjacent factory-applied coatings, overlapping by approximately 50 mm; see Figure 8.6.

Heat shrink sleeves are manufactured from radiation-crosslinked polyethylene tape, provided with a self-priming thixotropic sealant. The total thickness is 1.5–2.5 mm, with a backing tape of 0.6–1.0 mm. The application is similar to the above, except that there is only one cigarette wrap, which is made into a closed tube by a closure patch, placed at the top of the pipe. For onshore application, where the girth weld is performed in one operation, the heat shrink sleeve may be supplied as a closed tube, which is placed on one of the adjacent pipe joints prior to welding. Shrinking on to the joint is carried out using the yellow flame of a gas torch, applying the heat from the centre of the joint area and outwards. A typical shrinking ratio is 20–25% and, after shrinking, the overlap to the factory-applied coating should be at least 50 mm.

The difficulties in crosslinking polypropylene have recently been overcome. This has paved the way for the development of PP heat shrink sleeve

Figure 8.6 Application of tape wrap to field joint (photo: DONG A/S)

systems with higher corrosion, temperature and mechanical resistance. Using a liquid epoxy primer and a PP adhesive a temperature rating of 130°C can be achieved.

If the field joint is located between concrete coated pipe joints the gap is normally filled with a suitable material to produce a reasonably continuous pipe string. As the field joints typically constitute 6% of the pipeline length the infill material should not be too light so that the submerged weight of the pipeline is not significantly reduced; normally a minimum density of 2000 kg/m³ is specified.

The traditional infill choice is marine mastic, which is made from bitumen, limestone filler and gravel aggregate, with a typical density of 2100 kg/m³. The mixture is prepared in kettles, and poured into steel sheet shutters at a temperature of approximately 190°C. Cooling by water spray reduces the time before the field joint can resist the pressure of the rollers on the laybarge stinger. The form, which is typically 1.2 m long, is made from 0.7 mm steel sheet material, wrapped around the joint, fastened by steel straps on to the adjacent concrete, and provided with a filling aperture at the top. Figure 8.7 shows a completed field joint leaving the laybarge over the stinger. Note the spray nozzles cooling the marine mastic. The steel forms are left on the pipeline, but authorities may prohibit

Figure 8.7   Completed field joint going over the stinger

disposable formwork to avoid snagging of fishing gear and littering of the seabed when the straps eventually rust away.

Environmental authorities are also becoming less enthusiastic about marine mastic material, and dumping at sea is normally not allowed, which means that any excess or overheated mastic must be brought to shore for disposal. The preparation of the mastic in gas-fire kettles onboard a laybarge also constitutes a health and safety hazard, so alternatives have been developed.

For offshore application a prime concern is the application time, as the total time allotted for the coating of a field joint is 4–7 minutes. Various systems based on polyurethane (in foam or solid form) or other polymeric materials, such as isocyanates, with aggregates added for weight, are on the market. The rapidly curing mixture is injected into an annulus formed by reusable shutters, and application times of 3–6 minutes are reported. Rapid-setting concretes have also been tried, but in spite of successful onshore trial testing the offshore perform- ance has been disappointing.

Field joints prepared onshore, where time is of less concern, are normally infilled with traditional or rapid-setting concrete, cast in reusable formwork.

A special problem is posed by insulated pipelines, where the field joint areas must also be insulated. This may be done by means of pre-formed foam shells, or by injecting PU foam under a permanent shutter, strapped to the adjacent insulated joints. Water-tightness is ensured by cold tape wrapping, and the gap between adjacent concrete coatings is infilled as described above.

8.3.5 *Mechanical connections*

For offshore assembly, when pipe strings are joined to other pipe strings, or to risers, valve assemblies, etc., mechanical connections may be used as alternatives to welding, whether hyperbaric or above-water. Shrink fitted sleeves that use memory shape metal have been developed, as have axially slit, bolted sleeves and flanged couplings.

The most common type of mechanical connection is the flange joint, where flanges, bolts and gaskets are specified according to the pipeline pressure, type of external corrosion protection, and transported medium. Other possibilities include proprietary coupling devices; see Section 8.6.4.

8.4 Pipelaying

8.4.1 *General*

By far the most common – in terms of mileage – method of pipeline installation is by laybarge, where the pipeline is produced offshore by welding individual pipe joints into a pipe string, which is paid out from the laybarge to the seabed. Depending upon the shape of the suspended pipe we talk about S-lay or J-lay. The individual pipe joints, provided with the appropriate coating and anodes, are delivered by pipe supply vessels, the pipe being transferred by cranes mounted on the laybarge.

Smaller diameter pipe strings may also be fabricated onshore and spooled on to a reel, which is then installed by a reel barge. The two methods may be combined at the simultaneous laying of a large and a small diameter pipeline, with the latter piggy-backed on the former.

Pipelaying may be initiated at a shore approach (see Section 8.6.3) or at an intermediate offshore point. In the latter case the laybarge may pick up a previously laid pipe string or it may start pipelaying with an initiation head, connected to a dead man anchor or to start piles. A dead man anchor is a high holding anchor with a chain, laid down on the seabed to provide the required lay tension. In congested areas, however, it may not be feasible to place an anchor chain on the seabed, and instead start piles may be installed, using a subsea hammer. The tubular steel piles may be 15–20 m long, and are driven in until only the attachment wire is above the seabed, to avoid the need for subsequent removal. The laying of the pipe string is terminated with a lay-down head.

In the absence of lateral pipe supports (see Section 8.2.3) the maximum horizontal curvature that can be achieved during pipelaying depends on the seabed friction and the lay tension, as described in Section 6.3.6. It is recommended that one have a straight section of a length at least equal to the water depth between two curves in opposite directions, and both at initiation and at termination the pipeline should be straight for a distance of 500–1000 m. The lateral lay tolerance is typically ± 10 m, but it may be reduced to ± 2.5 m, although this is difficult to achieve in curves.

In most cases the laybarge is moored to eight or twelve anchors, and moves forward by pulling on the anchor cables. The anchors are relocated by tugboats, which together with the supply ships, survey boats and possibly diving support vessels constitute the laybarge spread. Some modern laybarges are provided with dynamic positioning (DP), and can keep station by powerful thrusters. However, DP is most suited for deep water, where the suspended pipe string is sufficiently flexible to absorb minor displacements at the surface without buckling.

8.4.2 S-lay

A laybarge is a floating factory where the pipe joints are welded on to the pipeline as it is installed. From the laybarge the pipeline describes an S-curve to the seabed. In the upper part (the overbend) the curvature is controlled by the laybarge stinger, a steel structure protruding from the stern of the vessel, that supports the pipeline on rollers. The curvature in the lower part (the sagbend) is controlled by lay tension transferred to the pipeline by tension machines gripping the pipe string on the laybarge. A typical pipelay configuration is illustrated in Figure 8.8.

The coated pipe joints are transported to the laybarge on pipe supply ships; see Figure 8.9, and stored on deck. Some major laybarges have double jointing facilities, implying that two 12.2 m pipe joints are welded together, usually by automatic submerged arc welding (see Section 8.3.2) before they are transferred to the end of the pipe string (the firing line), and welded on to the pipeline. The field joint coating, however, is carried out just before the stinger, usually at two stations working in parallel. To save time the welding on the firing line is carried out at a number of stations and as the weld is completed the pipeline goes into the tensioners, which are equipped with rolling tracks to allow movement of the pipeline whilst under tension, see Figure 8.10.

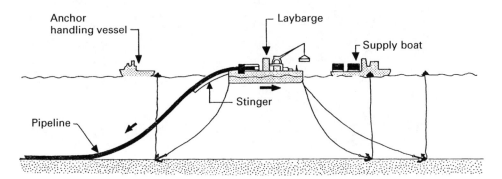

Figure 8.8 Typical S-lay configuration

Figure 8.9   Coated pipe joints on pipe supply ship being offloaded to the laybarge

Figure 8.10   Completed field weld entering laybarge tensioner (photo: DONG A/S)

At the last one or two stations the field joint coating is performed, and the laybarge advances one pipe joint (or two joints, in the case of double jointing and two field joint stations), moving under the pipeline, which goes out over the stinger. The lay rate is highly dependent upon pipe size, welding conditions, etc., but under optimal conditions a daily production (working 24 hr) of 4–5 km is not unusual.

8.4.3 *J-lay*

S-lay is feasible in water depths of up to approximately 700 m. At larger depths the weight of the suspended pipe string makes it impossible to maintain a stinger-supported overbend, and J-lay becomes an option. As the name implies, the discerning feature is that the pipe string enters the water in a vertical or nearly vertical position. This eliminates the firing line, which means that girth welding and field joint coating must take place in one or at most two stations only. Thus traditional welding procedures are too time consuming, and sophisticated methods such as friction welding, electron beam welding or laser welding are used. Even so, double jointing, or even triple or quadruple jointing, is essential to sustain a reasonable lay rate. J-lay barges are therefore equipped with a high tower to support the two to four pipe joints while they are being added to the pipe string.

Since there is no horizontal stinger there is no need for the pipe to enter the water at the stern of the vessel, which instead may have a mid-ships moonpool. Alternatively, a drillship may be converted to J-laying.

Using J-lay, pipelines may be installed in water depths exceeding 2000 m. At such depths dynamic positioning is the only feasible method of keeping station, but otherwise the operations, including abandonment and recovery procedures, are similar to those for S-lay.

8.4.4 *Reeling*

A reel barge is similar to a cable installation vessel. The pipe string is unwound from a vertically or horizontally mounted reel of diameter up to 30 m, pulled through a straightening device, and leaves the vessel over the stern. As the pipe enters the water at a steep angle the requirements to stinger support are minimal, and the sagbend is controlled by lay tension imparted by the reel. Obviously there is a limit to the diameter of the pipe that can be spooled, and currently the maximum feasible size for reel barge installation is 16″, although some examples of 18″ pipe reeling are reported. The wall thickness of a reeled pipeline will normally be governed by the strains imposed during installation, and depend upon the diameter of the reel. The pipeline cannot be provided with concrete weight coating, but adequate negative buoyancy is ensured by the heavy wall thickness, at least in deep waters which are not subject to much wave and current action. Indeed, reeling is most suitable for pipelines subjected to high pressures or large water depths, where the thick steel wall demanded by the installation method is also required to resist the functional or environmental loads.

The pipe string is manufactured onshore and spooled onto the reel. The individual string length depends upon the diameter, for a 12″ pipeline it may reach 10 km. Such spools are too heavy to be transferred offshore, so the reel barge will have to return to base to load a new string. In the meantime the pipeline is temporarily abandoned on the seabed, to be retrieved and pipelaying resumed as described in Section 8.4.6.

Once the new spool is joined to the pipeline the reeling only has to stop for the installation of sacrificial anodes at the spacing determined by the design. Bracelet anode half-shells for offshore mounting may be hinged together at one side to minimise the amount of field welding. On the other hand, this can reduce the clamping force, causing the anode to be sheared off during laying or trenching. The electrical cable connections are welded on to doubler plates; see Section 7.4.1.

8.4.5 *Piggy-back installation*

The need may arise for two or more pipelines to be installed along the same route, particularly in the case of interfield pipelines between platforms. Typically, these could be a larger hydrocarbon pipeline and smaller diameter service lines for injection water, lift gas or corrosion inhibitor. There may also be cables for power or communication, but at any rate it is economically advantageous to carry out the laying operations simultaneously. The laybarge may have room for two firing lines, but typically the secondary lines would be fabricated onshore and wound onto spools, whereas the main pipeline could be suitable for reeling or S-lay. Some S-lay barges may be equipped with a smaller stinger on the side from which a secondary pipeline can be paid out during the laying of the main line, but it is more common to clamp the smaller lines on to the larger in piggy-back fashion. For this purpose specially designed saddles are used, usually elastomer blocks strapped to the main pipeline by means of aramid fibre (Kevlar) or stainless steel bands; see Figure 8.11.

The piggy-back saddles may accommodate several smaller pipelines or cables, depending upon the individual sizes. As the main pipeline is being laid, the secondary lines are unreeled and clamped to the pipeline as is goes over the stinger. In addition to the mechanical attachment by the clamps the secondary pipelines must be electrically connected to the main pipeline. In order to be covered by the cathodic protection, the sacrificial anodes are designed to include the smaller pipelines as well. Alternatively, the piggy-back will also have to be fitted with anodes. Two pipelines of similar size may also be clamped together using elastomer or rubber spacers, and laid in one operation.

Figure 8.11 Laying of small (gas lift) pipeline, piggy-backed on larger (concrete coated) pipeline

8.4.6 *In-line components*

Large components such as valves, tees and wyes (see Sections 6.8 and 7.9) are typically installed in connection with riser bases, subsea manifolds, or other structures on the seabed, but they may also be installed in-line, i.e. by the lay vessel. Laydown heads, which may be provided with temporary pig traps, are in the same category. In this case, the component in question is welded into an assembly of the same length as a pipe joint, i.e. 12.2 m; see Figure 7.27. To ensure passage through laybarge tensioners it may, however, be necessary to perform some assembly welds offshore.

The longitudinal lay tolerance, i.e. the accuracy with which a particular point on the pipe string can be placed on the seabed, is typically ± 2.5 m, thus the target boxes for the initiation head and the termination head will be 5 m $\times$ 5 m. However, for a pipelay vessel moving in steps of 12.2 m another ± 6.1 m should be added to the longitudinal installation tolerance of an in-line component. In the case of the pipelay termination the target box accuracy is achieved by cutting off the last pipe joint before the termination head is welded on, but this is difficult to do for an in-line component without upsetting the welding sequence in the firing line.

8.4.7 *Abandonment and recovery*

Like all other offshore operations, pipelaying is weather dependent, and the tolerance depends upon the type and size of the pipelaying vessel and the supporting

Figure 8.12 Buckle detector being inserted into pipeline on laybarge (photo: DONG A/S)

spread. At a certain sea state it becomes impossible to add more pipe to the string, which is then kept under constant tension by the tensioners. Pipelaying will also have to be suspended if the weather prevents the tugboats from relocating the anchors, or the supply vessels from docking at the laybarge to transfer pipe or other essential supplies. If the movements of the laybarge become so large that they endanger the integrity of the pipeline the pipe string will have to be temporarily abandoned. A laydown head with an attached cable is welded on to the pipe string, which is lowered to the seabed under tension. If the laybarge is forced to abandon the site to seek shelter, the cable is attached to a buoy, for later retrieval. At the return of calm weather the pipe string is winched aboard the laybarge, secured by the tensioners, the laydown head removed, and pipelaying resumed.

The above abandonment and recovery operations are fairly routine, but they may also be invoked in the case of major mishaps. Weather induced movements or faulty manoeuvring of the laybarge may cause excessive bending of the pipe, resulting in buckling of the pipe wall. As a safeguard the laybarge may be equipped with a buckle detector, i.e. a gauging device that sounds an alarm if any reduction of the pipe internal diameter is measured; see Figure 8.12. The buckle detector is tethered inside the pipeline, trailing the sagbend touch-down point. If a buckle is detected the laybarge is backed up, retrieving the pipeline, and the affected joints are cut out. The same is the case if inspection of the radiographs or ultrasonic records identifies an unacceptable girth weld, which has gone over the stinger.

The resulting delays are manageable. A much more serious situation arises if a buckle results in a leak in the pipe, a so-called wet buckle. The pipeline must then be quickly lowered on to the seabed, or it may simply snap under its own weight. Recovery of a waterfilled pipeline may not be possible without further buckling, in which case the affected pipe string must first be dewatered. This entails the subsea installation of a pig receiver with dewatering valves (pipeline recovery tool), and possibly also a pig launcher at the other end of the pipe string, if it is located on the seabed. As a contingency against such eventualities the initiation head may be provided with dewatering pigs and valves for pressurisation with air.

A typical wet buckle contingency (WBC) procedure would comprise the following steps.

(1) Cut off and remove the damaged pipeline section.
(2) Hook up water pumping spread on a surface vessel to the initiation head.
(3) Displace the first pre-installed wet buckle pig with raw or inhibited seawater to remove sediment from the pipeline.
(4) Install the pipeline recovery tool at the cut end of the pipeline.
(5) Dewater the pipeline with compressed air from the surface vessel, using the second pre-installed wet buckle pig.
(6) Recover the pipeline to the pipelay vessel using the pipeline recovery tool.
(7) Depressurise the pipeline and resume pipelaying.

8.5 Towing, pulling and directional drilling

8.5.1 *General*

In installation by towing the pipe string is manufactured onshore, complete with field joint coating and sacrificial anodes, and towed to its final position. To control the pipe string at least two tugboats are normally deployed: one leading and one trailing. Depending on the level in the water, it is customary to distinguish between on-bottom tow (the pipe string dragging on the seabed), off-bottom tow (the depth controlled by hanging chains) and surface or below-surface tow (the depth controlled by floating pontoons). The various configurations are illustrated in Figure 8.13.

In controlled depth tow the pipe string is floating freely in the water column, the depth being controlled by the buoyancy of the string and the speed of the towing. This method is particularly suitable for a number of pipe strings and/or cables in a common carrier pipe (pipeline bundle). Several pipelines (or bundles) may also be joined on a common sled, and towed together.

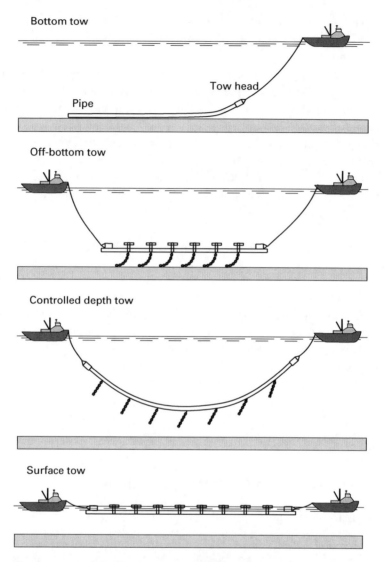

Figure 8.13 Pipe string towing configurations

If the pipe strings are dragged on the seabed by stationary winches we talk about bottom pulling, which is a common installation method for minor crossings and shore approaches. An alternative method in these cases is directional drilling, in which the string is pulled through a hole in the ground made through the soil by a rock drill.

8.5.2 *On-Bottom towing and pulling*

In on-bottom towing the pipe string is dragged over the seabed along a towing route, which must be carefully planned and surveyed to avoid damage to the pipeline. Obviously the coating must be abrasion resistant, and on-bottom towing is typically used for concrete coated pipe. Pipe strings have been successfully towed for hundreds of kilometres in the North Sea.

A particular variant is bottom pull, where the pipe string is pulled over the seabed by stationary winches, normally in a pre-dredged trench. This is a typical installation method for minor crossings, where the pipeline is welded up at one shore and pulled across by a winch on the opposite shore. The pulling speed is controlled by hold-back tension supplied by a winch attached by cable to the trailing end of the pipeline. To reduce seabed friction the pipe may be provided with buoyancy elements. Bottom pull is also the preferred method for shore approach construction; see Section 8.6.3.

8.5.3 *Off-Bottom, surface and controlled depth towing*

If the pipe is buoyant in the empty condition it is possible to tow the pipe string suspended in the water. The largest degree of control is achieved by off-bottom tow, which requires that the pipe string (including any buoyancy elements) is only slightly lighter than water. The distance above the sea bottom can then be controlled by chains attached to the pipe string, and dragging over the seabed. Should the pipe lift up, the added weight of the suspended chains will pull it down again.

In surface tow the pipe string is attached to pontoons, and floats on the surface of the sea. In this position the pipe is vulnerable to damage from waves, and the method requires sheltered waters or at least a calm weather window. Some protection may be obtained by suspending the pipe string from the pontoons, a method referred to as near-surface tow.

The most sophisticated towing method is controlled depth tow. This requires that the pipe string is trimmed to a slightly negative buoyancy, and the level in the water is controlled by the speed of the tugboats. In this way the pipe string can be kept well clear of wave action as well as of seabed hazards.

Pipeline bundles (see below), which can be made buoyant by the large sleeve pipe, are particularly adapted to controlled depth towing. For single pipe strings the method is less suitable because buoyancy elements would need to be added for flotation, particularly in the case of gas pipelines. The hydrodynamic stability criterion involves the pipeline in the operational condition on the seabed and, as the gas adds little weight, the empty pipe string will normally be too heavy for anything but bottom tow.

8.5.4 *Pipeline bundles*

An assembly of pipe strings enclosed in a common sleeve pipe is called a bundle. This, in addition to pipelines, also may include cables and umbilicals. The individual strings are fabricated to the required length, joined by appropriate spacers,

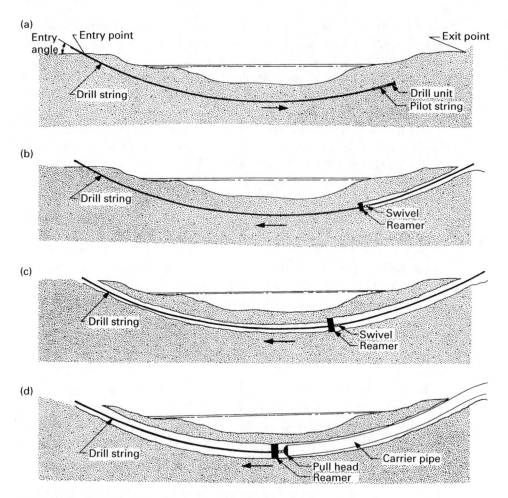

Figure 8.14 Principle of horizontal directional drilling

and threaded into the sleeve pipe, which is then sealed at the ends. The remaining space inside the sleeve pipe is filled with non-corrosive gas, inhibited water or thermal insulation material (see Section 6.6.5), whereas the sleeve pipe is provided with the required coating and sacrificial anodes for mechanical and corrosion protection. The pipeline bundle is installed by towing or pulling, as described above.

8.5.5 *Directional drilling*

Horizontal directional drilling (HDD) is a technique used in hydrocarbon exploration and production whereby the drill bit at the end of the originally vertical drill string is diverted sideways to an eventually horizontal direction, which allows the tapping of a large and shallow reservoir area from a single production platform. In the context of pipeline installation the term is used to designate an installation method in which the prefabricated pipe string is pulled through a hole in the ground made by a directed drill string. The method is illustrated in Figure 8.14.

At least one end of the pipeline has to be onshore, where the drill rig is placed, and a pilot string is inserted into the ground. The drill bit is hydraulically powered by bentonite drilling mud fed through the pilot string. The bentonite mud transports the soil away and fills the hole behind the drill head, preventing it from collapsing. The drill head is connected to the non-rotating pilot string by a swivel. The diameter of the cutting head is larger than that of the pilot string, which is encased by a drill string, and additional lengths of pilot string and drill pipe are added as the drill bit advances through the soil.

When the cutting head emerges at the exit point it is removed, and the pilot string is withdrawn through the drill pipe. A reamer is then attached to the drill string, which is pulled back through the hole, a wash pipe being attached behind the reamer. In the process the hole is enlarged by the reamer and, if necessary, the process is repeated with larger reamers. When the hole is sufficiently large to accommodate the pipe string, it is attached to the wash pipe and pulled through the hole with a reamer attached to the pull head as a precautionary measure.

Typical diameters would be 63 mm (2½″) for the pilot string and 125 mm (5″) for the drill string, with reamers of 350 mm (14″), 600 mm (24″) and 900 mm (36″). The latter will leave a hole that is sufficiently large to accommodate a 30″ pipeline. However, it is also possible to complete a 10″ pilot hole in one pass.

The route of the pilot string is determined by the entry angle and by the design of the drilling unit. The cutting head includes a hydraulic motor that uses the energy of the circulating drilling mud to rotate the bit. The cutting head is mounted on a bent transition unit (bent sub), the angle of which determines the curvature of the pilot hole, and forms the transition to the non-rotating pilot string. Any deviation from the prescribed path is corrected by rotating the pilot string, thus forcing the drilling unit into a revised direction. In this way the drill can be made to exit within a few metres from a target point located several kilometres away. If the exit point is unacceptable the pilot string is withdrawn a certain distance and the route corrected.

Determination of the current position of the cutting head is accomplished by one or more of the following devices:

- a pendulum providing inclination with the horizontal;
- a single shot survey camera providing tool face inclination and compass bearing;
- a plumb bob arrangement providing inclination;
- a triangulation system using sonar stations providing azimuth.

The success of the directional drilling method depends on the soil conditions, fairly uniform clay being the most appropriate. If the cutting head should hit a large boulder the string is retracted, and the hole redirected to avoid the obstruction. To avoid damage to the anti-corrosion coating as the pipe string is pulled through the ground, the coating must be abrasion-resistant; 3–4 mm polypropylene is a typical choice. Alternatively, a dual powder FBE system (see Section 7.4.5) can be used, or a conventional fusion bonded epoxy coating may be protected by a layer of polymer epoxy concrete or similar. Concrete weight coating is obviously not needed, as the pipeline is deeply embedded in the soil.

Directional drilling does not involve any activities between the entry point and the exit point, and is therefore a preferred method for crossing heavily built-up or environmentally sensitive areas.

A variation on the pull-back method described above is the large diameter forward thrust method. Instead of a drill pipe and pilot hole this procedure uses a cutting head mounted directly at the forward end of the pipeline and rotated by a hydraulic motor. The cutting head is a conical shaped tool with teeth embedded in the body, placed at the lead end of drilling unit containing the hydraulic motor, along with monitors for speed and direction, and a positive displacement pump. The unit is divided into two sections by a flexible accordion-type articulated coupling, making it possible to alter the direction of the unit by up to 3° either vertically or horizontally. The pipe string is made up as the drilling proceeds, by welding pipe joints behind the drilling unit. Thus the pipeline is drilled directly from the entry to the exit point.

8.6   Risers, shore approaches and tie-ins

8.6.1   *General*

Pipelines run between onshore terminals, platforms and/or subsea structures. A pulled or directionally drilled crossing is installed from shore to shore in one operation, but in all other cases the pipeline will consist of one or several pipe strings, which must be connected to one another and/or to termination points.

Each end of a given pipeline is located either onshore or offshore at an exploration or production facility. If the offshore installation is a platform a vertical connection called a riser is needed between the subsea pipeline and the topside equipment. The landfall of the pipeline is called the shore approach. It is normally constructed by pulling or directional drilling, but tunnelling or pipe jacking may also be used.

Tie-ins are the offshore connections between pipe strings, risers and shore approaches. The land-based connection to an onshore pipeline is outside the scope of this book.

8.6.2   *Riser installation*

In most cases the riser forms an integral part of the pipeline design, the solution depending on a number of project specific factors, such as the dimension of the riser, the product in the pipeline, the environmental parameters, whether the riser platform is new or existing, etc. The riser is fixed directly to the platform steel jacket by means of riser clamps, usually provided with elastomer linings. If the riser cannot be located inside the jacket structure a riser guard may be provided at the sea level as a protection against docking vessels or floating debris.

For greater protection, the riser may be installed in a caisson, which can be sized to accommodate several pipeline risers. The caisson may be dry and sealed at the top and bottom, the former seals being flexible to allow temperature movements. Also if the caisson is waterfilled it should be closed to prevent an exchange of water. The available oxygen will then be quickly consumed, obviating any corrosion even without the addition of inhibitors to the water. The caisson is attached to the jacket in the same way as a single riser.

At new platform installations the risers and/or the riser caissons are normally manufactured at an onshore fabrication yard together with the steel jacket. The activities offshore are limited to adjusting the level of the riser so the connection to the pipeline can be made. The pipeline tie-in is often made by a flanged spool piece, which fits the distance between the pipeline end flange and the riser bottom flange. To accommodate longitudinal movements of the pipeline end due to temperature and pressure variations the spool piece may include an expansion offset; see Section 7.9.2.

At existing platforms the installation of the riser is somewhat more complex. To minimise offshore work the riser is made out of prefabricated pieces that are as large as possible. The installation offshore takes place by crane, and the hook up to other installations may be made by flanged or welded connections. The connection to the pipeline is often made the same way as for pre-installed risers. For small diameter pipelines it is possible to connect the riser to the pipeline above water, and install the entire assembly using a barge-mounted crane, a method known as stalking.

At large water depths the riser may simply be a normal pipe string, which is installed as a catenary and connected to the surface installation and to the pipeline on the seabed. For small diameter pipelines, the riser section of the pipeline may be pulled from the seabed through a caisson fixed to the platform. Such caissons are referred to as J-tubes or I-tubes, the former being provided with a horizontal bend at seabed level. Alternatively, a flexible riser may be used; see Chapter 10.

8.6.3   *Shore approach construction*

Construction methods for shore approaches can be grouped under the following headings:

* bottom shore pull;
* bottom offshore pull;

- horizontal drilling shore pull;
- horizontal drilling offshore pull;
- tunnelling or pipe jacking offshore.

The most common method of installation is bottom pull, either to or from the shore. The pipe is pulled in a pre-dredged trench through the surf zone to a point above the high water mark. The depth of the trench should be sufficient to ensure that the pipeline is not exposed by seasonal or long-term variations of the seabed profile.

In the horizontal drilling version a pilot hole is drilled to a pre-dredged trench at the marine exit point. A crane barge with supporting equipment to handle drill pipe and hole openers (reamers) is positioned offshore, and the drill string is pulled on to the crane barge. A number of hole opening passes are carried out, until the drilled hole is sufficiently large to accommodate the pipeline, and the crane barge is then replaced by a laybarge or a reel barge.

Horizontal directional drilling is performed to overcome particular problems such as physical features at the landfall (sand dunes, cliffs, railways, roads, environmental restrictions, etc.). The geology at the landfall may favour horizontal directional drilling, and this method is often used to minimise disruption or ecological impact during construction of the pipeline landfall. Tunnelling may be used on rocky coastlines where adequate pipe protection can be difficult to achieve in the surf zone. Pipe jacking as a tunnelling method has been used on a few occasions in order to establish a pipeline route through an environmentally sensitive area.

Shore pull

At the prepared onshore site the pulling station is installed, usually consisting of two linear winches connected to a hold-back anchor, which may be a sheet pile wall. The winch cables have a sheave arrangement connected to the pull cable, which has been pulled in from the laybarge (or reel barge) stationed offshore, at the mouth of the trench. On the barge the pull cable is connected to a pull head, which is welded on to the first pipe joint, and the pipeline is pulled ashore as it is produced or paid out from the barge. A typical layout of the landfall site is shown in Figure 8.15, and Figure 8.16 shows the pull head emerging from the sea.

For horizontal drilling, the pipeline is connected to the wash pipe string, and pulled into the drilled hole from the barge.

Offshore pull

A pipe-stringing site is set up onshore, and the landfall pipe is welded into one string. A laybarge or reel barge is positioned at the pre-dredged trench or at the horizontally drilled exit hole on the seabed. Using the winches on the barge, the already prepared pipe string is pulled through the trench or the drill hole on to the barge, from where pipelaying or reeling is continued.

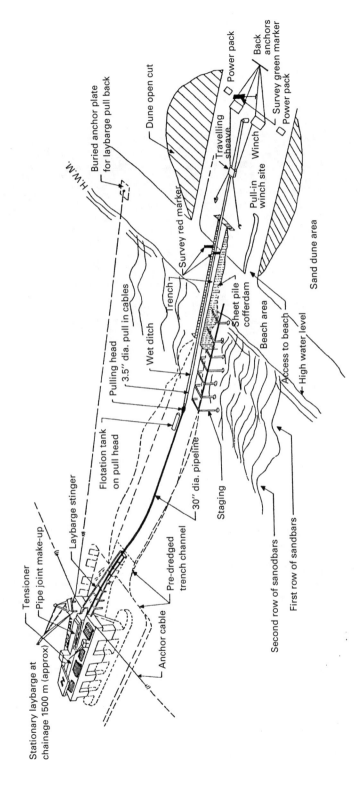

Figure 8.15 Typical landfall site layout

Figure 8.16 Pull head emerging from the sea at shore pull (photo: DONG A/S)

8.6.4 *Tie-in*

As mentioned above, the completion of a pipeline normally involves the joining of several pipe strings, at risers, at shore approaches or at intermediate points along the route. In some cases a pipelaying vessel may simply pick up the pipeline, remove the laydown head, and continue pipelaying, but often logistics dictate that tie-in of pipe sections already on the seabed must be performed.

In shallow water, such as close to the shore approach, it may be feasible to lift up the two pipe ends by barge-mounted davits, clamp them in position, and execute the tie-in weld above water. After weld acceptance and field joint coating the pipeline is lowered on to the seabed, the barge moving sidewards to avoid overstressing of the pipe steel; see Figure 8.17.

At greater water depths the tie-in must be carried out on the seabed by hyperbaric welding; see Figure 8.18. A purpose-built habitat with alignment clamps is lowered over the two ends, the water is displaced by compressed air, and welder-divers in saturation perform the cutting, bevelling, alignment, welding and field joint coating, as described in Sections 8.3.2 and 8.3.4 above albeit omitting any field joint infilling. If the pipeline is flooded the end must be closed off with a waterstop, consisting of a bladder deployed during flooding, and inflated to seal the pipe at a predetermined position.

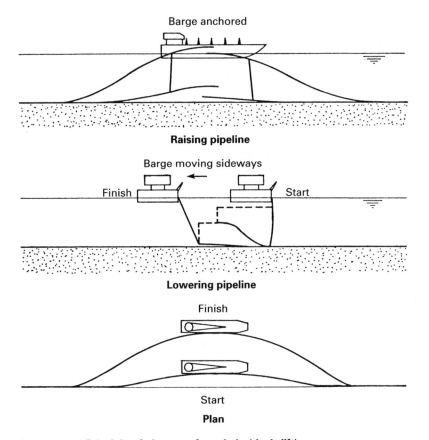

Barge anchored

Raising pipeline

Barge moving sideways

Finish   Start

Lowering pipeline

Finish

Start

Plan

Figure 8.17   Principle of above-surface tie-in (davit lift)

Alternatively the two pipe ends may have been provided with weld neck flanges, capped by blind flanges. Hyperbaric welding is then replaced by a bolting together of the two flanges, after removal of the blind flanges. For flooded pipelines the alignment and bolting together may be carried out by divers without the use of a dry habitat. If the pipeline is dry the end may be sealed off with inflatable waterstops.

Saturation diving is feasible up to a water depth of approximately 180 m, and in deeper waters it is necessary to resort to diverless methods, using mechanical connectors. The tie-in is carried out by means of a proprietary, ROV-operated tie-in system (RTS); there are a number of makes on the market.

A typical RTS is a combined pull-in and connection tool, which docks on to a tie-in porch located on one of the ends to be joined, normally a structure; see Figure 11.1. The pull-in is done with a winch, whereas the connection (stroking) is accomplished by hydraulic cylinders. One of the hubs to be connected (the inboard hub) is fitted to the tie-in porch, and the other (the outboard hub) is located on the pipeline or spool that is being pulled in. Any blind flanges (pressure caps) are removed by the tool prior to tie-in. The pipe bore will thus be

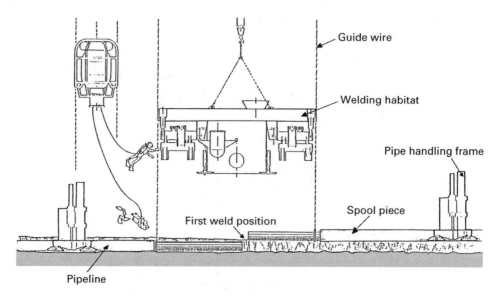

Figure 8.18   Principle of hyperbaric welding tie-in

flooded after the tie-in operation, but spools or flowlines in corrosion resistant alloys that do not tolerate seawater may be filled with fresh water and sealed off with gel plugs or similar.

8.6.5   *Branch line connection*

Spur lines are connected to the main pipeline by means of tees or wyes, a typical example being the tee assembly shown in Figure 7.27. A long trunkline may have several such in-line tees pre-installed to facilitate the tie-in of pipelines from future satellite fields.

Hot-tapping is a technique for tie-in to a live pipeline. Proprietary, remotely operated tools have been developed for drilling into a pipeline and connecting a branch line without the need to empty the pipeline first. A compromise between a classical hot-tap operation and the pre-installed tee described above is a hot-tap tee; see Figure 8.19. This component is easier to install than the tee assembly, as it does not have to be closed off with an isolation valve, and it is provided with a hub for receiving the hot-tap tool and for the future tie-in operation.

8.7   Trenching and backfilling

8.7.1   *General*

Permanent installation of the pipeline below the natural seabed is called trenching. The objectives of trenching are:

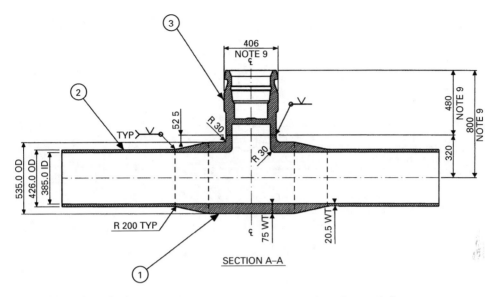

Figure 8.19 Cross-section of a hot-tap tee piece with hub for a future tie-in (sample drawing)

- to protect the pipeline from hydrodynamic forces;
- to protect the pipeline against mechanical damage;
- to eliminate or reduce free spans;
- to prevent upheaval buckling;
- to increase thermal insulation of the pipeline.

Seasonal or long-term variability of the seabed bathymetry (e.g. sand waves) may require trenching to ensure the permanent integrity of the pipeline.

In some cases, particularly for towed or pulled pipe strings, the pipeline may be placed in a pre-dredged trench, but in most cases post-trenching is required to lower the pipeline into the seabed. Post-trenching methods include:

- water-jetting;
- mechanical cutting;
- ploughing.

If required, the trench may be backfilled by ploughing or rock dumping, but often natural backfilling will take place. On sandy seabeds natural self-lowering of the pipeline may occur, obviating the need for trenching to ensure stability.

8.7.2 *Jetting and cutting*

Water-jetting is performed by means of a jet sled, which is riding on the pipeline, guided by rollers at the top and the sides of the pipe; see Figure 8.20. The jet sled is pulled by a trench barge, which also delivers the compressed water that is

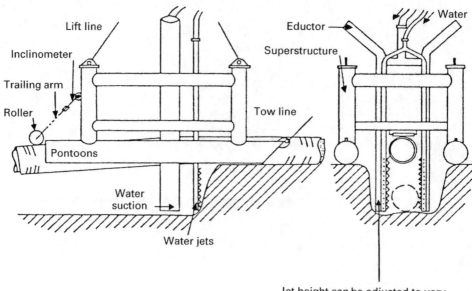

Figure 8.20   Typical water-jetting sled

ejected through nozzles at each side of the pipeline. The nozzles may be arrayed vertically or mounted on inclined swords of adjustable length. The water lique-fies and displaces the seabed soil, leaving a trench into which the pipeline sinks. This process is aided if the trenching is carried out when the pipeline is water-filled, but it must be documented that this will not lead to overstressing of the linepipe steel. Several passes of the jet sled may be necessary to achieve the specified distance from the natural seabed to the top of the pipe.

Jetting is most efficient in soft or sandy seabeds, but can also be used in cohesive soils with shear strengths of up to approximately 100 kPa. In stiff clay mechanical cutting is preferred. The trenching tool rides on the pipeline, being pulled by a trench barge, but instead of water nozzles it is equipped with cutting heads, which excavate a V-shaped ditch in the seabed.

Whether jetting or cutting, the trenching vehicle may also be self-propelled on tracks or skids, and remotely operated. This obviates the need for a pull cable, but power and control will still have to be delivered from a surface vessel. In deep waters only remotely operated trenching systems are used, and a recent NORSOK standard U-102 *ROV services* specifies requirements for such systems.

8.7.3   *Ploughing*

The trench may also be excavated by a plough, which is clamped around the pipeline is such a way that the shears displace the soil from under the pipeline,

Figure 8.21   Trenching plough being deployed from a trenching vessel (photo: DONG A/S)

depositing it in walls alongside the trench. The plough is supported on skid beams, and is pulled by a surface vessel, being guided along the pipeline by rollers. Figure 8.21 shows a trenching plough being deployed from the trenching vessel.

Trenching ploughs have been developed that can deal with all soil types, from soft mud to shale or limestone. Depending upon the specified trench depth and the nature of the soil more than one pass of the plough may be necessary.

8.7.4   *Artificial backfilling*

Whether the pipeline is trenched or not, a soil cover may be needed to protect the pipe from mechanical damage from fishing gear or minor anchors, or to prevent upheaval buckling. If natural backfilling cannot be relied upon the cover will have to be established by dumping rock material on the pipeline; see Section 8.2.3.

The mere trenching of the pipeline will normally be sufficient to ensure hydrodynamic stability, and if a soil cover is necessary for long-term integrity this will often be achieved by natural backfilling. Indeed, in sandy soils where the trenching is performed by water-jetting, some soil will be suspended rather than displaced, and will settle around the pipeline at the bottom of the trench. Seasonal storms will ensure a complete filling of the trench. When the trenching is carried out with a pipeline plough, backfilling may be achieved in the same operation by providing the plough with two sets of shears. The first pair opens up the trench, depositing the displaced soil alongside. The second pair then scoops the soil back on top of the pipeline. The distance between the two pairs must be sufficient to allow the pipeline to describe an S-curve from the top of the seabed to the bottom of the trench, thus the method is only suitable for small diameter pipelines or umbilicals. For large diameter lines the plough would have to be too long to be practical, so instead it is necessary to perform two passes with a plough: the first for trenching, the second for backfilling.

If hydrodynamic stability and not mechanical protection is the only concern, trenching of the pipeline may be obviated by reliance upon self-burial, i.e. the tendency of any heavy object to sink into a sandy seabed due to the scouring action of waves and current. An attractive option is to leave the pipeline water-filled on the seabed during a winter season, and the following spring post-trench only those sections that have not embedded themselves sufficiently to be stable when the pipe is emptied. The propensity for self-burial may be enhanced by attaching spoilers in the form of longitudinal fins to the pipeline, but experience with this method is limited to test sections. The deposition of sediment on top of the pipeline can also be promoted by artificial seaweed, i.e. fronds of buoyant polymer attached to a net anchored on the seabed.

8.7.5 *Protective covers*

Localised covering of the pipeline may be required, for example, to protect against dropped objects at platforms, or to prevent scour in the vicinity of platform legs or other structures on the seabed. In such cases a structural cover is an alternative to rock dumping. The engineered covering may be constituted by structural concrete elements placed over the pipeline, built up by flexible mattresses manufactured from geotextile, bitumen and aggregate or from interlocked concrete blocks. Also sand bags or grout bags placed by divers may be used.

A particular need for protective covering occurs if the pipeline has to cross existing pipelines or cables on the seabed. If the line to be crossed is not provided with sufficient soil cover seabed intervention is necessary. The most common method is the dumping of rock berms (see Section 8.2.3), but other possibilities include mattresses (bitumen or concrete) or specially designed cover structures. The latter two options may be required for small service lines or umbilicals, which might be damaged by rock dumping.

8.8 Pre-commissioning

8.8.1 *General*

Pre-commissioning, also known as RFO (ready for operation), covers all activities from performance of the acceptance pressure test, normally part of the scope

Figure 8.22   Typical de-watering pig (photo: DONG A/S)

for the installation contractor, up to filling the competed pipeline with the product, and the commencement of product transportation.

The activities of de-watering and drying are particularly important for gas pipelines, because any remaining water may react with the gas to form hydrocarbon hydrates, which can obstruct the flow, in particular hindering the proper functioning of valves. A typical de-watering pig is shown in Figure 8.22.

Smaller flowlines or service lines, normally designed with suitable reserve for corrosion allowance, can be commissioned simply by running product behind a separation pig.

8.8.2   Flooding and hydrotesting

When all construction activities (e.g. pipelaying, tie-in, trenching, crossing construction and artificial backfilling) have been carried out, the final integrity of the installed pipeline is documented by hydrostatic testing. This requires that the pipeline be water-filled, and seawater is normally used for this purpose. It would be advantageous, from a corrosion standpoint, to use de-aerated water, but de-aeration is not practical for the quantities needed for long, large bore transmission pipelines. Seawater is pumped into the pipelines through a simple water winning arrangement that includes filtering and sometimes treatment of the seawater with oxygen scavengers and biocides to prevent internal corrosion of the linepipe steel; see Section 6.6.3. The oxygen scavenger removes the oxygen that may fuel corrosion, and the biocide prevents the growth of anaerobic bacteria. When determining the additives consideration shall be given to the possible trapping of water in valve cavities or branch piping, and to any effects on valve and seal materials.

A typical oxygen scavenger is sodium bisulphite ($NaHSO_3$), a dosage of 65 mg/l (ppm) being required for an oxygen concentration of 10 ppm. A common biocide is glutaraldehyde at an active concentration of 50–75 mg/l (ppm). As glutaraldehyde reacts with sodium bisulphite the oxygen scavenger should be given a few minutes reaction time before the biocide is added, or alternatively an over dosage must be used. Some commercially available sodium bisulphites are combined with a catalyst, which may reduce the requirement for time delay or over dosage.

An alternative biocide is sodium hydroxide (NaOH), also known as caustic soda or lye. To reach a pH of 10.3, which is lethal to most organisms, a dosage of $0.4-0.6$ l/m^3 of 30% NaOH is needed. However, the use of lye will result in large amounts of precipitated carbonates and hydroxides, which may impede the function of valves, and form calcarious deposits that are not easily removed from the pipe wall.

At any rate, the use of chemicals in pipelines often attracts the attention of authorities, and for pipelines having landfall it also attracts possible interested parties near these. It is therefore tempting to design the pipeline and test the pipe materials in such a way that hydrotesting can be avoided completely. In practice this approach has yet to be implemented, even though, for example, the DNV code does permit it for deep water applications. In this context it may be noted that subsea tie-in operations may also require temporary flooding of the pipeline; see Section 8.6.4.

The requirements of chemical treatment are very much dependent upon the residence time of the test water in the pipeline, and many owners operate with a time limit of 60 days as a requirement for obviating biocides. As discussed in Section 6.6.3, the general corrosion of carbon steel pipes exposed to stagnant seawater is very limited. If the amount of corrosion products is considered uncomfortable it is possible to add an oxygen scavenger, which is widely viewed as considerably less environmentally foreign than biocide.

For corrosion resistant alloys, whether as solid pipe or as clad or lined pipe, it may be worth performing hydrotesting using chloride-free fresh water to prevent chloride induced pitting corrosion. The water may be treated by oxygen scavenger to avoid corrosion products. Biocide is not required as fresh water contains no sulphates, thus SRBs do not present a problem. Precautions should be taken to prevent the ingress of seawater during installation and tie-in operations. Should seawater enter the system it should be removed by freshwater flushing as soon as possible, and within a maximum of 7 days in the case of 13% Cr SMSS. Duplex or austenitic stainless steels are more tolerant, and a residence time of 20 days may be allowed. Some CRAs (e.g. superduplex or molybdenum alloys) are resistant to cold seawater, thus no special precautions need to be taken for water temperatures below 15°C, the 20 day limit applying for warmer waters.

Hydrostatic testing comprises a strength test as well as a leak test, and is carried out by pressurising the water to the specified leak test pressure, which is maintained for the specified holding period. The holding period should take into account that there needs to be time for temperature variation stabilisation. The holding period should not be taken as less than 24 hours after stabilisation has been documented. During the holding period the pressure is closely monitored, and any pressure drop that cannot be ascribed to variations in atmospheric pressure, water levels or seawater temperature signals a leak, which must then be localised. To facilitate leak detection the test water can be mixed with a powerful dye or a hydrocarbon tracer, which can be sensed by a 'sniffer' fish that is towed along the pipeline.

Owing to environmental concerns, recent designs of transmission lines tend to avoid dye, or at least minimise its use. To achieve this, dye sticks are mounted at critical locations, such as valves or tie-in points. Dye sticks or dye applied as a paint can be inserted by divers just prior to tie-in operations. The dye stick can, again for environmental consent reasoning, be made of what is popularly labelled 'invisible' dye, which is fluorescent and visible only to a diver carrying an inspection tool.

Should a leak occur, which has been known to happen, it normally takes the form of a violent rupture, which is easily localised even if the pipeline has been trenched and backfilled. As an example, Figure 8.23 shows the failure of a pipeline that ruptured during hydrostatic testing. The pipe had been subjected to repeated gouging, but hydrogen embrittlement was a contributing factor; see Section 6.7.3.

If a visual survey does not locate the failure it is possible to launch a 'pinger' pig, which can be tracked acoustically until it stops at the rupture. Alternative means of location include the use of magnets or radioactive sources. Another type of leak is the so-called pinhole leak, that is next to impossible to detect. The pipeline section to be hydrotested may be isolated by subsea valves, and for these a maximum leak rate across the seats is specified. This magnitude of leakage, however, is not detectable during pressure testing.

8.8.3 *Gauging*

Even if the pipeline survives the hydrotesting it should be ascertained that there are no dents in the linepipe wall, which could induce failure in the long term, or obstruct the passage of cleaning and batching pigs. For this purpose gauging and caliper pigs are propelled through the pipeline during water filling. The caliper pig is a so-called intelligent pig, equipped with sensors that measure the internal diameter at a number of points around the circumference, and it is not normally used during construction. The device is sufficiently sensitive to pick up the

Figure 8.23   Rupture of gouged pipeline during hydrostatic testing

individual girth welds, and produces a chart showing the average bore against the distance travelled. In this way any anomaly can be located for diver inspection and cut out if necessary.

The gauging pig is normally a simple aluminium plate that is recovered and inspected during construction activities. As a successful gauging run is often a contractual interface, and certainly a key component in the insurance of the pipeline, the contractor will try to perform this as early as possible. More than one gauge plate is often propelled through the line, particularly when the installation including pre-commissioning is split between a number of contracts.

According to DNV OS-F101 the diameter of the gauge plate should be 97% of the nominal pipe ID, but a smaller plate diameter may well be adequate in order to take account of weld root penetration and misalignment, particularly for small bore pipelines. The gauging pig is normally incorporated in one of the pig trains used to water fill and clean the pipeline interior (see Figure 8.24 below), after which the test water is displaced from the pipeline.

8.8.4   *Cleaning*

During and after water filling the pipeline interior should be cleaned. Apart from any inflatable waterstops left over from tie-in operations, which of course must be pushed out, an amazing amount of debris may have accumulated in a long trunkline: not only discarded welding rods and cigarette packs, but also gloves,

Figure 8.24   Brush pig being removed from the pig trap (photo: DONG A/S)

hard hats, etc., as well as the occasional lost buckle detector. If the pipeline is not internally coated several pigging runs are necessary to remove rust and mill scale, which might otherwise clog up valves during operation. The combined rust and mill scale layer at the interior surfaces may be less than 0.1 mm thick, but the accumulated volume over a 100 km long 30″ pipeline can amount to several cubic metres. The use of an internal flow coating combined with modern manufacturing of the pipelines reduces this refuse significantly.

The cleaning trains include both brush pigs and swapping pigs, the latter removing any brushes that may have broken off. The pig trains are normally propelled by the treated seawater pumped in for the purpose of the hydrotest, but further cleaning by running brush and swapping pigs in air may take place during and after de-watering. Figure 8.24 shows a brush pig being removed from the pig trap.

In Figure 8.25 a typical flooding, cleaning, and gauging pig train is shown. Note that the length of the train is 900 m. When designing the pig train considerations into bi-directionality and simple travelling length must be conducted. Pigs can travel several hundred kilometres, but then the wear and tear of the first pigs must be taken into account.

As seen in Figure 8.25, the cleaning operation may be facilitated by gel-plug technology. A gel is a plastic fluid capable of picking up loose and loosely

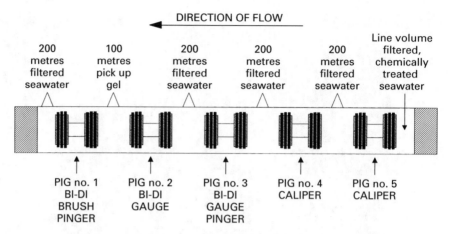

Figure 8.25 Example of pig train used for flooding, cleaning and gauging

adhering solids. The gel slug is inserted into the pipeline, followed by an appropriately designed scraper pig. The train should consist of more scraper pigs to collect any gel slipping by the pig driving the gel. The plastic fluid moves through the pipeline in a manner known as plug flow. The central part of the slug moves as a semi-solid plug with little exchange of material with the fluid making up the annular flow region adjacent to the pipe wall, which moves slower than the mean velocity of the total gel plug. The core of the gel in front of the mechanical pig, moving faster than the gel on the outside closer to the wall, creates a tractor action, pulling and lifting the debris-laden gel away from the front of the pig and into the gel plug. The debris, which would remain in front of the pig in a conventional operation, is thus picked up and eventually distributed throughout the length of the slug. Gels can be produced with a range of viscosities, including solid gel pigs, capable of removing wax or paraffin deposits.

8.8.5 *De-watering*

The de-watering operation must be planned with a view towards the disposal of the water, particularly if it is treated with corrosion inhibitors, as dumping in coastal areas is not likely to be acceptable; see Section 8.8.2 above. For minor crossings the water may be collected in tanker trucks and transported for onshore disposal, but for interfield pipelines and larger trunklines offshore discharge is the only option. The water is discharged through a diffuser head to ensure a dilution to a concentration that is not harmful to marine life, a fact which must be documented by environmental impact studies. Thus for a pipeline terminating at a landfall the de-watering must be performed from shore, and if this is not feasible a temporary outfall pipeline must be constructed so that the water can be discharged at sea.

These problems can be mitigated by using untreated test water. As discussed in Section 8.8.2, the amount of dissolved oxygen is insufficient to cause significant corrosion, and the duration of waterfilling may be too short for anaerobic bacteria (SRB) to flourish. In this case the test water may be unsightly, particularly if the pipe has not been internally cleaned and coated, but it does not constitute an environmental hazard.

De-watering of pipeline runs is carried out by means of air-propelled pig trains during or after cleaning; see above. Non-piggable items, such as crossover spools, may be provided with drain panels and vent panels, allowing expulsion of the water by compressed air and filling with methanol or similar. Even so, valve bodies or by-pass lines may trap some water, which must be accounted for in the subsequent drying operation.

8.8.6 *Drying*

If the pipeline is to be used for natural gas a complete drying is necessary, as any residual water may react with the gas to form hydrates, which may obstruct the flow and impair the proper functioning of the valves. The presence of water will also make any impurities of hydrogen sulphide (H_2S) and carbon dioxide (CO_2) highly corrosive. To dry the pipeline the following methods may be used, alone or in combination:

- methanol (or glycol) swapping;
- hot air drying;
- vacuum drying.

In the swapping method a batch of methanol or tri-ethylene glycol (TEG) is enclosed between pigs and propelled through the pipeline by compressed air. Residual water will be dissolved in the hygroscopic substance, leaving a film that is mostly methanol or glycol. It can be mentioned that this swapping is somewhat debatable, the argument against swapping being that by the time the swapping occurs corrosion has taken place and any residual water in the pipe will have dried up anyway.

An alternative procedure, which combines cleaning and drying in one operation, is gel pigging, as described above. Modern gel-forming agents can produce gels from an array of liquid components. By incorporating gels based on hygroscopic fluids, such as methanol, into the cleaning train the water is removed along with the debris.

Hot air drying exploits the ability of hot air to hold a large amount of water in the form of vapour, whereas vacuum drying relies upon the lowering of the boiling point of water at low pressures. For long trunklines, vacuum pumps will need to work for several days or weeks to decrease the pipeline pressure below a few millibar. At that pressure the water dewpoint will be less than −30°C, indicating adequate dryness of the pipeline. To limit the time, vacuum drying is often used as the last step, i.e. after most of the water has been removed by swapping or gel pigging.

8.8.7 *Nitrogen purging*

To prevent any internal corrosion between pre-commissioning and operation it is customary to fill the pipeline with a non-corrosive gas, such as nitrogen with a typical purity greater than 95% (i.e. 95% N_2, 5% atmospheric gases). However, nitrogen with 5% atmospheric air should not be considered non-corrosive if there is free water in the pipe. The amount of oxygen is large enough to cause severe corrosion over a prolonged period of time, and if aerobic/anaerobic bacteria are also present the pitting tendency may be large. To achieve non-corrosive conditions the nitrogen should constitute more than 99.98% of the gas. If the pipes are completely clean and dry there is no need to fill the pipe with nitrogen or any other form of non-corrosive gas.

For a vacuum dried gas pipeline the nitrogen is simply let in, in other cases the air in the pipeline is displaced by nitrogen, a process known as purging. Liquid nitrogen is vaporised through heat exchangers and injected into the pipeline. To guarantee a low level of oxygen the volume of injected nitrogen should be approximately twice the volume of the pipeline.

Nitrogen is introduced at the upstream end of the pipeline with a −50°C dew point or lower, at a controlled rate to prevent over-compression and subsequent re-condensation of water. Dew point control is critical, and the infill rate and controlling pressure should be determined to ensure that at no time is the dew point above −20°C. Whilst the initial purge is performed, regular monitoring of the oxygen content of the atmosphere in the vicinity of the discharge point should take place.

Nitrogen should be discharged and the dew point monitored until the separation pig has been received, during which time the nitrogen dew point should be −20°C or drier at atmospheric pressure at the outlet end of the pipeline. The pipeline should then be packed with nitrogen to a final pressure of typically 5 barg. The selection of the required nitrogen overpressure is a matter for interpretation and discussion. As nitrogen is expensive a slight overpressure of 1.1 barg has been used. The argument for increasing the pressure, as used recently for a gas pipeline in moderate water depth, is that pinhole leaks will result in gas outflow rather than water ingress.

When completed the pipeline is in what would normally be the final 'handover' condition. The installation or pre-commissioning contractor will then demobilise. Product filling for operation will be the responsibility of the pipeline owner and operator; see Chapter 11.

Chapter 9
Control and documentation

9.1 Introduction

Apart from the nuclear industry not many, if indeed any, civil, technical design areas are subjected to such elaborate quality assurance procedures as the offshore industry, and this also applies to the design of marine pipelines. Moreover, authorities often require the pipeline to be certified by an independent third party, such as Det Norske Veritas (DNV), Bureau Veritas, Lloyds of London, etc. The certification comprises all major activities during pipeline design, fabrication and installation.

The following sections outline some reasonable requirements for the control and documentation of the project. The latter comprise the documents for operation (DFO) or life cycle information (LCI), which constitute the basis for all future operation, maintenance, repair or modifications of the pipeline system. Frequent reference is made to the DNV OS-F101 and related recommendations, which is only one way of specifying LCI requirements, but it may serve as an illustrative example.

The receiving party is termed the Owner, and the executing party is termed the Contractor. The latter may include sub-contractors and suppliers, whether or not these have a direct contractual relationship with the Owner. Many of the subsequent sections are in the form of check-lists, which may contain terminology that is not readily understandable by the non-specialist, and reference is made to the Glossary, where most of the terms are defined.

9.2 General requirements

9.2.1 *General*

Each design activity should be carefully described, and consensus obtained within the organisation regarding its execution. This may be achieved by means of regular meetings and/or through the use of formal 'design briefs', which are short write-ups of the proposed methodology to be applied during detailed design. Similarly, all supplies and construction activities shall be performed in accordance with approved specifications and procedures.

9.2.2 *Quality management*

All suppliers, contractors, designers and operators should establish and maintain formal systems to control the issue, production, registration, modification and filing of all documentation. Document management is normally performed under a quality assurance system in accordance with the ISO 9000 series, and environmental management is handled in accordance with the ISO 14000 series. Similar formal systems should be put in place with respect to occupational health and safety.

It is common to integrate the Quality Assurance (QA) and the health, safety and environmental protection (HSE) issues into the same management system. The important point is that there should be clear allocation of responsibilities regarding reporting, remedial action and documentation.

9.2.3 *Document formats*

The parties should adopt a standard format for all documents with a standard title block for all project drawings and standard document sizes conforming to ISO standards, and front sheets, check sheets, progress reporting, etc. in an agreed format.

Dimensions and all other units of measurement should be in accordance with the SI system. However, it is customary to denote nominal pipe diameters in inches and pressure ratings in bars. When a pipe dimension is stated it must be clear whether it is the internal diameter (ID) or the external diameter (OD).

Each page of a document should be numbered in a manner that indicates the total number of pages within the document (e.g. 1 of 8, 2 of 8, etc.). The document number and the current revision should also be indicated on each page.

Hard copy of all documents, manuals, reports, plans, procedures, drawings and final documentation should be backed up with an electronic copy on CD, or e-mail in the case of short documents. Table 9.1 gives an example of format requirements given to a Contractor.

Any documents that cannot be held electronically should be of sufficiently high quality to allow scanning without loss of information.

9.2.4 *Communication*

All communication, whether written or spoken, is normally in the English language. It is important to ensure the linguistic skills of all participants and operators. Thus all site supervisors should be fluent in English.

All formal correspondence should be addressed and directed between dedicated company representatives. Formal correspondence is deemed to include letters, faxes, e-mails and minutes of meetings. Such items should have unique reference numbers.

Table 9.1   Example of document format specification

Document type	No. of hard copy sets	No. of electronic copies	Electronic format
Installation manuals	1	2	• Adobe Acrobat pdf file
Monthly reports	1	1	• Adobe Acrobat pdf file
Intermediate monthly reports	0	1	• Adobe Acrobat pdf file
Plans and procedures	1	1	• Adobe Acrobat pdf file
Field reports	1	1	• Word/Excel and AutoCAD dwg files
As-built documentation	1	4	• Adobe Acrobat pdf file • AutoCAD 2000 dwg files • Key data sets in source program (Excel, ASCII, etc.)

Oral or non-formal communications of instructions or information in connection with the works should be confirmed in writing using formal correspondence.

9.2.5   *Document register*

The Contractor should maintain a document register that typically fulfils the following requirements:

- presents the controlling dates for each document to be produced by the Contractor;
- presents the current status of Contractor documents;
- carries a new revision number whenever changes are made;
- highlights, by means of a summary, where changes have been made from the previous revision.

The document register should be updated whenever documentation is being produced, and could fit into monthly progress reports.

9.2.6   *Documentation review and acceptance*

All documents produced by or on behalf of the Contractor should be subject to a review, to verify compliance with statutory or contract requirements and good engineering practice, and should be approved for issue by authorised personnel.

A procedure for achieving acceptance of documentation could comprise:

(1)   first draft of completed document submitted to the Owner;
(2)   Owner comments returned within 10 working days of receipt;

(3)   Owner comments incorporated by the Contractor and document resubmitted within 10 working days of receipt of comments;

(4)   document either accepted or returned by the Owner to the Contractor (revert to step 2).

The review status could be in accordance with the following.

Review Status A: No comments. The Contractor may proceed with the activities covered by the document. Signing off the front sheet as approved.

Review Status C: A document comment sheet is issued, indicating that the contractor may proceed, subject to the implementation of comments. The document shall be revised and reissued to obtain review status A.

Review Status R: A document comment sheet will be issued by the Owner, indicating that the document is not acceptable. The document will require major revision and re-issue.

9.2.7   *Nonconforming items*

Any object or occurrence that is at variance with agreed specifications or procedures should be documented. An appropriate format is a Nonconformity Report, containing the following information:

- identification of project and activity;
- identification of responsible personnel;
- identification of the affected item;
- description of nonconformity, including identification of the violated requirement with reference to the appropriate regulatory document;
- justification of the adopted technical decision (e.g. use as is, repair, scrap);
- description of any repair or remedial work;
- documentation of the appropriate approvals.

If suitable, the Nonconformity Report should also make reference to any corrective action taken to prevent any recurrence of the nonconformity.

9.3   Design

9.3.1   *Design basis*

A design basis for a pipeline system should be developed. The document could be a static design basis, e.g. prepared by the Owner, or it could be a live document, continuously updated whenever design information is received.

The document should contain:

- a full description of the pipeline system and interfaces, pressure regulating system and other safety issues, functional requirements, system life time, and similar key data;
- principles, design codes and methods for strength and in-place analysis;
- pressures, temperatures, fluid composition and all other relevant operational data;
- geotechnical data;
- meteo-marine information;
- topographical and bathymetrical conditions along the route.

See also Chapter 2.

9.3.2 *Design documentation*

The activities during the design phase should be suitably documented to enable independent verification by the Owner, the Authorities, and third parties, including a possible certifying agency.

The design should cover all relevant structural and environmental evaluations including:

- pipeline route;
- materials selection;
- pipe wall thickness;
- strength analysis under anticipated loading during installation, hydrotesting and operation, including installation loads, trawl interaction, forced displacement in trench transitions or over crossings, etc.;
- temperature and pressure profiles;
- pipeline expansion force behaviour, including upheaval buckling;
- risk assessment, quantitative or for selected risks;
- corrosion prediction and monitoring;
- pipeline stability and free span development and pipe response.

9.3.3 *Design drawings*

Design drawings should be prepared suitably to present the measures for fabrication and installation of the pipeline system. The drawing register should include:

- pipeline route drawings, including alignment sheets with seabed properties and water depths, existing or planned restrictions and obstructions, such as platforms, shipping lanes, lighthouses, cables and pipelines (live or abandoned); see Section 2.6.2;

- typical pipeline detailed drawings, including coating, field joints, anode fabrication and fastening;
- out-of-straightness and span rectification requirements;
- detailed drawings of crossing designs;
- detailed drawings of tie-in arrangements, including riser details;
- shore approach drawings.

9.4   Supply and fabrication

9.4.1   *General*

This section covers requirements to documentation and inspection related to the supply and fabrication activities:

- pipe production;
- anode manufacture;
- pipe coating, including anode attachment;
- other supplies.

9.4.2   *Pipe production*

Control of the linepipe production begins at the steel making facility, and the critical plate rolling operations; see Chapter 3. The operation at the pipe mill covers incoming plate materials through to finished and tested linepipe. Owing to the significant cost and the obvious importance to the finished pipeline, the description of pipe production is very detailed in the various existing codes. Most widely used is API 5L, which in Europe, however, is increasingly being replaced.

Pipes in Europe are typically manufactured according to ISO 3183-3 with various additional requirements as stipulated by the Owner. As in the case of many other pipeline topics, probably the most complete code coverage is found in EN 10204 and DNV OS-F101 Section 6 with subordinate Appendices A through E.

A concluding comment on the level of detail in the various manufacturing codes and standards could be that the fabrication of linepipe, however specialised a task, does not require the same elaboration of procedures and other project specific documents as the installation work.

Certification
Pipes are normally delivered with EN 10204-3.1C certificates or equivalent, stating product number, heat number, manufacturing process and delivery conditions, and containing all results from specified inspection, testing and measurements, etc. The 3.1C certificate, however, is not required by ISO or DNV, the

philosophy of the ISO certification system being that manufacturers can document the quality and independence of their in-house control.

Document requirements

ISO 3183-3 Annex B includes the following documentation requirements for all pipe:

- steel maker;
- steel making and casting techniques;
- aim cast chemistry and range;
- hydrostatic test procedure;
- non-destructive testing procedure for the pipe.

See ISO 3183-3 for specific additional requirements for welded or seamless pipe manufacture.

The information supplied according to ISO 3183-3 Annex B can be supplemented with the following information:

- calibration procedures;
- methods of identification/traceability;
- procedures for handling, stacking during storage, and securing pipe for transport;
- plans;
- results from the manufacturing procedure qualification tests.

For each pipe, information should be required on characteristics that may cause systematic defect signals on intelligent pig survey charts. This is not a standardised requirement, and may be hotly disputed by linepipe manufacturers, but it is useful to have documentation of the locations of features such as weld repairs or thickness reductions, for example those due to local grinding or pipe forming, even though the acceptance criteria are satisfied.

Documented properties include:

- residual magnetism;
- carbon equivalent (CEV, calculated from the chemical analysis);
- yield strength ($R_{t0.5}$);
- ultimate tensile strength (R_m);
- strength ratio ($R_{t0.5}/R_m$);
- elongation;
- hardness (sour service);
- charpy V values.

All test data should be presented as a series of histograms, with mean values and standard deviations.

Manufacturing procedure specification

The pipe mill and other fabrication yards should prepare manufacturing procedure specifications, and describe a qualification test to demonstrate that the proposed properties will indeed be achieved. The manufacturing procedure specification should pinpoint the document flow to validate that the specified properties are obtained during production.

Documentation prior to start of production

A listing of the documents to be submitted to the Owner prior to the start of production should include:

- manufacturer's quality manual;
- project quality plan;
- organisation of the work, with resumes of all key personnel;
- manufacturing procedure specification;
- manufacturing schedule;
- manufacturing procedures, including test requirements and acceptance criteria.

Documentation after production

Production control is documented by the pipe mill project quality plan, and continuously verified through submission of control documentation to the Owner's site representative.

The documentation of the finished product should include:

- all records from the manufacturing procedures, including test requirements and acceptance criteria;
- material certificates with complete records for chemical composition, heat treatment, strength properties, dimensions and weights;
- weld records;
- production test records such as visual inspection, dimensional control, NDT records, cut out samples, etc.;
- hydrostatic testing report (mill testing).

9.4.3 Anode manufacture

The operations at the anode production facility covers the following operations:

- receipt of insert, cable and alloy materials;
- fabrication of steel inserts;
- casting of anodes;
- connection of electrical cables;
- shipment of finished anode half bracelets.

Requirements for the manufacture and testing of anodes are given in NACE RP 0492-92 and DNV RP B401.

Certification
Anodes should normally be delivered with EN 10204-3.1B certificates or equivalent, stating product number, melt number, manufacturing process and delivery conditions, and containing all the results from the specified inspection, testing and measurements, etc.

Quality and environmental management
The anode supplier should document the implementation of a formalised quality and environmental management system, as well as health and safety procedures. A Quality Plan for the anode production should define inspection and testing methods, their frequency and acceptance criteria.

Manufacturing procedures
The anode supplier should provide an anode manual, containing all procedures and documentation related to the production activities, such as:

- detailed, dimensioned fabrication drawings;
- material certificates and chemical composition documents;
- insert welding procedures;
- anode casting procedures;
- inspection and testing procedures;
- certification and calibration certificates for testing equipment;
- description of anode handling, stacking and storage methods;
- description of anode transportation methods;
- methods of anode identification and traceability;
- details of document management system.

Documentation prior to start of production
A list of the documents to be submitted to the Owner prior to start of production should include:

- Quality Manual
- Quality Plan;
- quality system certificate;
- safety policy statement;
- manufacturing procedure specifications, as outlined above.

Documentation after production
Production control is documented by the anode production Quality Plan, and continuously verified through submission of control documentation to the Owner's site representative.

The documentation of the finished product should include:

- all records from the manufacturing procedures, including test requirements and acceptance criteria;
- material certificates with complete records for chemical composition, dimensions and weights;
- insert weld records;
- production test records such as visual inspection, dimensional control, NDT records, cut out samples, etc.;
- electrical continuity testing records of cable connections.

9.4.4 *Pipe coating*

The operations at the pipe coating facility cover some or all of the following operations:

- receipt of linepipe, anodes and coating materials;
- internal coating;
- external coating and insulation;
- anode bracelet attachment;
- shipment of finished coated pipe joints.

Requirements for the application and testing of pipe coatings may be found in DNV RP F106, and anode attachment is covered by DNV RP B401.

Certification
Coating is normally covered by EN 10204-3.1C certificates or equivalent, containing results of production testing and measurements, etc. Certificates must ensure full traceability of the coated item by reference to the product manufacturer's assigned number. Any identification marking that is covered by coating must be transferred.

Quality and environmental management
The pipe coater should implement a formalised quality and environmental management system, as well as health and safety procedures. A Quality Plan for the coating operation should define inspection and testing methods, their frequency and acceptance criteria.

Coating procedures
The pipe coater should provide a coating manual, containing all procedures and documentation related to the coating activities, such as:

- coating material certificates;
- cleaning and preheating procedures;

- coating application procedures;
- anode attachment procedures;
- inspection and testing procedures;
- certification and calibration certificates for testing equipment;
- description of pipe handling, stacking and storage methods;
- description of pipe transportation methods;
- coating and pipe repair procedures;
- methods of pipe identification and traceability;
- details of document management system.

Documentation prior to start of production

A listing of the documents to be submitted to the Owner prior to the start of coating operations should include:

- Quality Manual;
- Quality Plan;
- quality system certificate;
- safety policy statement;
- coating procedure specifications, as outlined above.

Documentation after production

Production control is documented by the pipe coating Quality Plan, and continuously verified through submission of control documentation to the Owner's site representative.

The documentation of the finished product should include:

- all records from the coating procedures, including test requirements and acceptance criteria, as well as records of repairs; and
- production test records such as visual inspection, dimensional control, weight control, etc.

9.4.5   *Other supplies*

Other supplies that are included in the pressure containment, such as flanges, tees, wyes, bends, instrument pipings and valves, should meet the same requirement to documentation of quality and properties as those valid for the linepipe material. In addition, valves and all other pieces of equipment that are necessary for the safe operation of the system shall be of approved design and fully documented quality and functionality. Operation and maintenance procedures for all subsea valves and actuators should be described in detail.

Documentation should also be included for all items related to internal coating (where applicable) and external corrosion protection, as well as to external bracings, supports and scour protection. The quality and properties of the coating materials should be documented by relevant material certificates and data sheets.

9.5 Installation

9.5.1 *General*

The engineering contractor and, subsequently, the various installation contractors should compile installation manuals, documenting the project specific installation procedures for undertaking the work.

The specifications and procedures should describe as clearly as practically possible the control and monitoring to be performed during installation, including clear directions on acceptance criteria.

Offshore inspection, representing the Owner and possibly the certifying agency, monitors the compliance to the best of their ability. The automatic readings generated during the installation work, e.g. recorded roller pressures, tensioner loads and automatic ultrasonic testing records, are important in assisting the monitoring of compliance with specifications.

9.5.2 *Installation manuals*

The Contractor should prepare all pre-construction documentation in the form of a series of installation manuals, e.g. in accordance with DNV OS-F101, section 9A 500.

An installation contract will typically be split into a number of work packages or scopes, and the Contractor should prepare an installation manual for each item, containing all relevant documentation required for that work scope. The installation manuals should be prepared, reviewed and accepted by the Owner prior to mobilisation.

It is essential that individual documents from an installation manual can be traced back to the acceptance status of the manual. Either all documents within a manual should have the same revision number as the manual, or an index should be included at the front of the manual, indicating which document revisions are covered by the manual.

The installation manual should include all documentation required to perform the installation, and clearly provide in tabular or graphical form all the information that the Owner's inspection personnel require to monitor and validate fully the execution of the work.

9.5.3 *Contractor's engineering and management*

As part of an installation contractor's engineering and management, a number of procedures and specifications should be submitted after the award of the contract, but prior to the start of construction works. These may include:

- job descriptions for key personnel;
- initial issue of document register;

- project quality plan;
- Level 3 installation schedule, giving breakdown of activities on a daily basis;
- work breakdown structure (WBS) and progress measurement procedure;
- cost plan;
- procedures, schedule and application plan for the acquisition of authorisations, licences and consents;
- detailed materials take off (MTO), including Contractor supplied items and consumables. The MTO shall include the quantity of pipe required for weld procedure qualification;
- materials control procedure;
- procedures and schedules for the Contractor's dealings with the Authorities regarding authorisations, licences and consents for which the Contractor is responsible;
- audit schedule;
- mobilisation schedule for main vessels;
- first issue of the health, safety and environmental protection (HSE) plan, including supporting procedures such as an HSE specification, to be updated as required throughout the work;
- materials safety data sheets for all hazardous materials to be used during the work;
- format and sample as-built alignment sheet drawing;
- format of as-built survey incident lists, incident codes and transverse profiles;
- details of accommodation, services and communications systems on board all vessels;
- format of weekly construction update report;
- as-built report index for each work scope.

9.5.4 *Installation procedures*

Depending upon the size and complexity of a pipeline project, a number of specifications will be required to be made by the Contractor, based on the Owner's specifications for the work.

To identify possible critical items or activities it may be beneficial to carry out systematic analysis of installation operations and equipment, so-called Hazard and Operability (HAZOP) studies and Failure Mode Effect Analyses (FMEA).

The documentation submitted for Owner review prior to the start of construction might include:

- health, safety and environmental specification;
- survey procedures detailing survey requirements, including equipment for the various stages of the work;
- installation procedures for all the construction methods applied, including monitoring and acceptance criteria, validation of equipment and vessels, qualification of operators such as welders, and all documents to be delivered;

- trenching specification;
- shore approach construction works procedures;
- subsea structures fabrication and installation procedures;
- pipe protection and seabed intervention procedure;
- hydrotest procedure;
- de-watering and drying procedure.

In the following sections a selected number of key work packages are detailed. Note that the listing cannot be generally applied, i.e. complete engineering of project specific requirements would normally be a part of the project execution.

Offshore pre-lay survey

An installation manual, including supporting specifications, for an offshore pre-lay survey would include the following:

- survey equipment and software details, specifications for key equipment;
- equipment calibration procedures and checklists;
- vessel acceptance trial procedures;
- manning levels;
- pre-lay survey procedures;
- tidal model;
- any special procedures for location of live cables;
- procedures for obstruction identification and removal;
- procedures for minor re-routing around obstructions;
- contingency procedures;
- details of all reporting formats;
- results of pre-lay HAZOP and associated FMEA studies performed.

Marine Pipelay

An installation manual, including supporting specifications, for marine pipelaying would include the following:

- general arrangement and elevation drawings of the laybarge, showing locations of all major equipment required for installing the pipeline; elevation drawings to include pipelay configuration drawings and details of the stinger for all configurations to be adopted during the work with sufficient information to enable review of roller heights and positions;
- a schedule of proposed tensioner settings for the entire pipeline;
- calibration of tensioner and constant tension winch;
- ranges of permissible roller reactions for critical rollers;
- verification of roller friction;
- predicted location of pipe catenary touchdown point, and prediction of suspended pipeline configuration;
- tension requirements for abandonment and recovery operations;

- testing and inspection of tensioning equipment;
- cable payout/position of laydown head during abandonment and recovery operations;
- assessment of station keeping performance in high currents;
- local buckling unity checks for the different stages of the installation and for each coating configuration, pipe wall thickness and water depth combination encountered;
- local buckling unity checks during retrieval of a dry buckled pipe from the maximum water depth along the route;
- local buckling unity checks during retrieval of a wet buckled pipe from the maximum water depth along the route;
- buckle detection system details;
- laybarge and anchor positioning procedures, anchor patterns and graphs;
- survey equipment and software details and matrix of key equipment;
- procedures for pipelay initiation, normal lay and laydown;
- contingency procedures;
- procedures for pipelay at crossings, including contingency procedures in the event that the pipeline is not bearing on the prepared supports or significant settlement takes place;
- DP operating procedures;
- diving manual;
- vessel and equipment acceptance trials programme;
- details of all QA reporting formats that are to be used;
- production welds test proposals;
- material handling and repair procedures;
- field joint coating procedures;
- wrapping tape/heat shrink sleeve data sheets;
- infill system data sheets;
- infill system handling procedure;
- results of pipelay HAZOP and associated FMEA studies performed.

Trenching and backfilling

An installation manual, including supporting specifications, for trenching and backfilling would include the following:

- specifications and general arrangement drawings of trenching/backfilling support vessel(s);
- specifications and drawings of trenching and backfilling equipment, including pipeline configuration during deployment, initiation and recovery of the trenching/backfilling machine, and the location of sensors/instrumentation;
- local buckling unity checks for all stages of trenching and for each coating configuration and pipe wall thickness combination encountered, consideration being given to a range of trencher attitudes if relevant to pipeline loading;
- allowable operating limits for instrumentation readings;

- trench stability calculations;
- design and operation calculations and proposed field trials for new or untried trenching or backfilling equipment;
- calculation of transition zone lengths;
- procedures for DP trials;
- equipment calibrations, including methodology for load cell calibration;
- vessel and equipment acceptance trials programme;
- vessel and trenching/backfilling equipment positioning procedures;
- survey equipment and software details and matrix of key equipment;
- limiting weather and sea state conditions for safe deployment, initiation and recovery of the trenching/backfilling equipment;
- procedures for all operations, including deployment and recovery, and including trenching at subsea services crossing locations;
- procedures for removal of obstructions during trenching;
- emergency/contingency procedures, in particular with respect to protection of third party pipelines/cables at crossings, and including temporary abandonment and recovery of equipment on the seabed;
- anticipated trenching and backfilling rates;
- procedures for monitoring trenching and backfilling performance and interpretation of instrumentation readings;
- results of trenching HAZOP and associated FMEA studies performed.

9.5.5 *Construction reporting*

Weekly construction updates

A typical time span between construction status reports is one week. As an example a weekly construction update report could include the following index:

- status of work undertaken/achieved during the previous 7 days;
- listing of the various major vessels operating in each specific area, including locations in geographical coordinates;
- location of guard vessels;
- pipe tracking records, collection of the daily reports of when pipe movements have taken place;
- copies of all inspection records, collection of the daily reports of inspection activities;
- listing of sections of the pipeline that have been laid, but remain untrenched, defined in geographical coordinates and overall length;
- location of any span heights in excess of 0.5 m in geographical coordinates;
- status of each cable/pipe crossing (i.e. prepared, laid over, trenched to, finally rock dumped);
- list of major equipment in use, and overall status of installation with respect to onshore sections at each end of the system;

- details of any factors that could influence the schedule;
- list of planned operations for the forthcoming 7 days.

Weather forecast

The Contractor should provide daily weather forecasts to the relevant project sites and offices. The source of the weather forecasts should be agreed between Contractor and Owner. The weather forecast should be updated twice daily, typically with the following forecast information:

- synoptic situation;
- gale warnings;
- wind speeds at 10 m and 50 m;
- wave height (significant, maximum and extreme) and period;
- swell height, direction and period;
- state of weather (e.g. rain, fair, etc.);
- visibility;
- air temperature.

All vessels should be provided with means of receiving weather data transmission. The Contractor should take action to minimise the impact of weather on the work, and prepare emergency procedures covering the actions to be undertaken should bad weather be forecast for any weather sensitive operation.

Daily construction reports

The Contractor should provide daily construction reports to the project offices. The report might have the following index:

- Vessel name
- Date and location
- Safety and environmental incidents
- Material movement status
- Work completed in the last 24 hours
- Downtime/Standby time with reasons
- Work planned for next 24 hours
- Fuel, lubricants and potable water used
- Helicopter movements and numbers of passengers in/out
- Details of any diving operations
- Current and forecast weather conditions and sea state summary
- Any deviations from terms and conditions given in authorisations, licences and consents
- Contractor comments
- Owner comments
- Persons on board, with changes

9.6 As-built documentation

9.6.1 *General*

The content of the as-built report should be developed by the Contractor during the progress of the work, and should include all items that are listed in the description of the individual work scopes.

The complete project documentation may be collated in a Design, Fabrication and Installation (DFI) resume. The objective of the DFI resume is to provide the information database for the operation of the pipeline, and to assist in the preparation of inspection and maintenance planning.

All design documents, supplier information, specifications, procedures, certificates, inspection reports, etc. are collated in the DFI resume, which may comprise a substantial number of volumes.

Computerised database systems that also function as management tools during operation of the pipeline (see Section 9.6.2), have been successfully used for several large projects.

During operation the pipeline will be surveyed at regular intervals. The documentation required to undertake the survey activities include:

- organisation chart of people responsible for the pipeline operation;
- as-built drawing base, possibly updated with previously recorded abnormalities and corrective measures, if any;
- production records, detailing pressure, temperature and flow characteristics;
- records from previous inspections.

9.6.2 *Pipeline management systems*

The storage, verification and retrieval of as-built information for major pipelines is greatly facilitated by a computerised pipeline management system. All production data from the pipe producer, anode manufacturer, pipe coater, and installation contractor are entered into a common database, providing on-line access to the information. To be effective the management system should be specified from the outset of the project, and it should be ensured that the data from the various parties are provided in compatible electronic formats. All information is recorded against the unique Pipe Mill Number that each pipe joint is assigned during linepipe production, but it should be possible to retrieve data through the following search keys:

- Pipe Mill Number;
- Linepipe steel heat number;
- Pipeline installation sequence number;
- Pipeline chainage on the seabed.

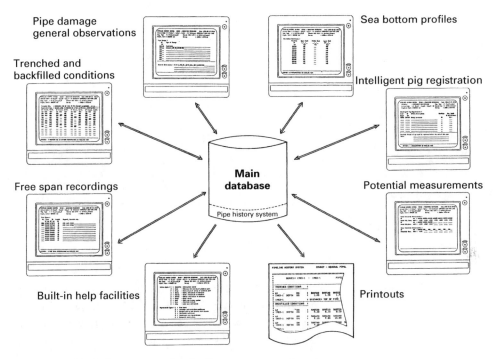

Pipe damage
general observations

Sea bottom profiles

Trenched and
backfilled conditions

Intelligent pig registration

Main database

Pipe history system

Free span recordings

Potential measurements

Built-in help facilities

Printouts

Figure 9.1 Typical structure of a pipeline management system

In the operational phase the management system can be supplied with inspection data. The pipeline may be divided into segments, against which, for example, the following information can be recorded; see Figure 9.1:

- length and location of free spans;
- trench and backfilling conditions;
- seabed and soil conditions;
- anode potentials;
- intelligent pigging records;
- damages or interesting seabed features.

The technology of management systems is developing rapidly, and modern versions tend to be web-based, which means that the interested parties have access to an intranet where all information is posted, and can be retrieved by a few mouse-clicks. In addition to the traditional databases, which essentially replace printed files, the system can store video clips, GIS data, etc.

Chapter 10
Flexible pipes

10.1 Introduction

10.1.1 *General*

Unbonded flexible pipes are an alternative to rigid steel flowlines and risers. In particular, the use of flexible pipes in connection with floating production systems (FPS) is an area that has been subject to a rapid growth since the beginning of the 1990s; see Figure 10.1. A major advantage of using the flexible

Figure 10.1   Floating production system

Figure 10.2   Unbonded flexible pipe

pipes is their ability to work under extreme dynamic conditions and their relatively good insulating and chemical compatibility properties compared with rigid carbon steel pipes. Furthermore, flexible pipes are used as tie-in jumpers due to their ability to function as expansion spools, and the jumpers can be installed without carrying out a detailed metrology survey. The flexible pipes are used for a multitude of functions, including production and export of hydrocarbon fluids, injection of water, gas and chemicals into an oil/gas reservoir, and service lines for wellheads.

Flexible pipes can be manufactured in long continuous lengths. Consequently, long flowlines can be installed without introducing intermediate joints, thus minimising the risk of leaking flange connections (flowlines with a continuous length of up to 8.5 km have been installed in the North Sea area). The present size range of flexible pipes is from 2″ to 19″. Note that whereas the nominal size of a rigid steel pipe normally refers to the outer diameter (OD), the size of a flexible pipe indicates the bore, i.e. the internal diameter (ID). The internal pressure rating is typically in the order of 70 to 700 bar (1000–10 000 psi) depending upon pipe size, water depth and function. Flexible pipes are presently used at fluid temperature up to 130°C.

Unbonded flexible pipes are custom designed, complex multi-layered structures, built from a number of helically wound metallic wires or strips combined with concentric layers of polymers, textiles, fabric tapes and lubricants. The number, type and sequence of the component layer depends on the specific design requirements. A typical conventional unbonded flexible pipe structure is shown in Figure 10.2, consisting of carcass, inner liner, pressure armour, tensile armour and outer sheath.

All the layers in the flexible pipe are terminated in an end fitting, which forms the transition between the pipe and the connector (e.g. a flange, clamp hub or weld joint). The end fitting is designed to secure each layer of the pipe fully so that the load transfer between the pipe and the connector is obtained whilst maintaining fluid tight integrity.

Conventional free hanging flexible pipes are limited to a water depth of approximately 2000 m, depending on pipe diameter and internal pressure. As the offshore oil industry continues to move into deeper waters, there is an increasing demand for the development and qualification of new technologies to enable this expansion to take place. Consequently, alternative flexible pipe designs are being developed, using composite materials to reduce the pipe weight and to increase the tension capacity.

10.1.2 *The world's first subsea flexible pipes*

The first subsea flexible pipe concept with a size of ID 3″ was secretly developed during World War II as part of the British war-time project PLUTO (Pipe Line Under The Ocean) for the transportation of fuel across the English Channel to Europe during the Normandy invasion. However, it was not until 1968 that the first commercial flexible pipe was installed between Iceland and the Vestmanna Islands, located 14 km off the coast of the mainland; see Figure 10.3. The incentive for developing this flexible pipe was provided by the urgency of supplying drinking water to the islands as their water reservoirs had become contaminated with sulphuric ashes from volcanic activities. A supplier of subsea cables, that had previously installed a power cable between the mainland and the islands, was encouraged to deliver and install an ID 4″ pipeline with a pressure rating of 70 bar. The driving force for adopting a flexible pipe technology was governed by the very uneven seabed topography at a water depth of 100 m.

10.1.3 *Unbonded flexible pipe specification*

Experience with offshore flexible pipes has been gained since the 1970s, when flexible pipes were manufactured for use in relatively calm environments, such as offshore Brazil. However, with the increasing demand for flexible pipes intended for the North Sea from the early 1980s, new industry guidelines were developed, reflecting the more severe environmental conditions applicable for this particular geographical area.

The initial specifications for flexible pipes were primarily derived from the manufacturers' experience of in-house testing programmes. Generally accepted industry standards were developed in the late 1980s as presented in the DNV *Guidelines for flexible pipes* (1987), and the API RP 17B *Recommended practice for flexible pipe* (1988).

Bureau Veritas was the first classification society to issue a certification scheme as described in *Non-bonded flexible steel pipes used as flow-lines*, which was published in 1990.

Figure 10.3   Installation of the first commercial subsea flexible pipe in 1968 by NKT's lay barge *Henry P. Lading*

In 1994 DNV issued *Rules for certification of flexible risers and pipes*, including utilisation factors for a variety of failure modes and load combinations. Also, design requirements concerning upheaval buckling of flexible pipes were presented for the first time in such a document.

In 1996 API issued the industry standard API Spec 17J entitled *Specification for unbonded flexible pipes*, covering design, materials, manufacturing, documentation and testing of unbonded flexible pipes. Furthermore, in 1998 a comprehensive update of the recommended practice API RP 17B was issued (revised again in 2002), addressing unbonded and bonded flexible pipes for onshore, subsea and marine applications. Bonded flexible pipes are dealt with in a separate API Specification 17K. At present, a new API Spec 17L industry standard is being developed to cover ancillary equipment for the flexible pipe.

Figure 10.4   8 km of ID 8″ flowline on a turntable

10.2   Flexible pipe structure

10.2.1   *General*

Unbonded flexible pipes are characterised by having a low bending stiffness combined with a high axial tensile stiffness. This is achieved by designing a composite pipe wall, using a combination of polymeric sealing layers and helically wound metallic profiles; see Figure 10.2. The high bending flexibility allows these pipes to be spooled on relatively small diameter cylinders. Thus, flexible pipes may be stored, transported and installed from dedicated offshore reels with a typical core diameter of 3–5 m and an outer diameter of 9–10 m.

The maximum length of pipe that can be spooled on an offshore reel would, for a typical 8″ pipe, be about 3 km. Should a reel be insufficient to accommodate the desired pipe section length, a turntable (or carousel) may be used; see Figure 10.4.

Normally, a flexible pipe is designed specifically for the intended application, and consequently it is not a typical off-the-shelf product. A variety of different layer combinations are possible for the composite pipe wall depending upon the application. To standardise this, API RP 17B defines the following three flexible pipe families:

(1)   Smooth bore pipe (product family I);
(2)   Rough bore pipe (product family II);
(3)   Rough bore reinforced pipe (product family III).

The build-up of the composite pipe wall for the above product families are specified in Table 10.1, showing the main structural layers. Other layers may be

Table 10.1   Classification of standard, unbonded flexible pipes (API RP 17B)

Main structural layer	Product family I (smooth bore)	Product family II (rough bore)	Product family III (rough bore)
Internal carcass		X	X
Inner liner	X	X	X
Pressure armour	X		X
Intermediate sheath	X*		
Tensile armour	X	X[†]	X
Outer sheath	X	X	X

* The use of an intermediate sheath is optional.
[†] The cross-wound tensile armour may be applied with a lay angle close to 55° to balance radial and axial loads.

applied, such as manufacturing aid layers (e.g. tapes), thermal insulation, and anti-wear layers for dynamic applications (see Section 10.2.8). In the following the main structural layers used in flexible pipes will be described in more detail.

10.2.2   *Internal carcass*

The innermost layer in family II and III pipes (rough bore pipes) is the carcass, which is not a leak proof structure. The purpose of the internal carcass is to provide the pipe with radial support to resist external loads (crushing and hydro-static pressure), and to serve as a mechanical protection of the polymeric inner liner, e.g. against erosion by solid fluid particles (typically sand) and during pigging of the pipe. The collapse capacity of the carcass should be determined by assuming that the external hydrostatic pressure is acting directly on the inner liner tube, to take into account that the pipe annulus may be flooded (e.g. due to damaged outer sheath).

The conventional carcass is fabricated by cold forming a flat stainless steel strip into an interlocking structure as shown in Figure 10.5. Typical materials used for the carcass include AISI 304L and AISI 316L stainless steel grades. However, other grades such as duplex and super duplex steel materials may be used as well, depending upon the required corrosion resistance. Figure 10.6 also shows an alternative carcass structure, specifically developed for deep and ultra-deep water applications (2000–3000 m).

10.2.3   *Inner liner*

The polymeric inner liner, which serves as the leak tight barrier in the flexible pipe, is made from the extrusion of specific grades of polyethylene (e.g. HDPE, XLPE), polyamide (e.g. PA11) and polyvinylidene flouride (PVDF) materials. The relatively limited choice of polymeric materials available for flexible pipes is governed by the requirements of large-scale extrusion (diameter, wall thickness

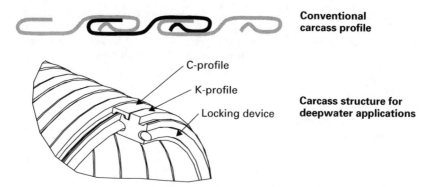

Figure 10.5 Typical carcass structures used in unbonded flexible pipes

and pipe length) and the problems associated with time dependent degradation of the material in the extrusion equipment. Normally, the inner liner is extruded in one layer only, but two- and three-layer extrusions are made where special sacrificial layers or thermal barriers are required towards the carcass and/or the pressure armour. Special attention should be given to possible pressure build-up between the layers of the multi-layer inner liner structures, caused by gas permeation from the pipe bore. This may result in collapse failure of the carcass when the pipe bore is de-pressurised. Typical maximum allowable design temperatures for the polymeric inner liner materials are 65°C (HDPE), 95°C (PA11 and XLPE) and 130°C (PVDF).

10.2.4 *Pressure armour*

The pressure armour layer provides the required radial strength capacity to resist internal pressure loads and external crushing and hydrostatic pressure loads. Furthermore, the pressure armour can be designed to support the carcass structure thus increasing the collapse capacity of the pipe, due to, for example, the carcass failing at higher buckling modes. The resulting interaction between the pressure and carcass armour layers due to external radial loads depends upon the relative strength capacities of the profiles in the two armour layers (e.g. yield strength and moment of inertia).

The pressure armour layer is made of helically wound metallic wires or strips with a high lay angle relative to the longitudinal pipe axis. Examples of typical interlocking wire profiles are shown in Figure 10.6. The wires are normally plastically pre-formed during the manufacture to fit the outer diameter of the polymeric inner liner with a minimum of residual bending stresses. The armour wires are normally made of low-alloyed carbon steel grades with high yield strength, typically in the order of 800–1000 MPa. However, steel grades with lower yield strength may be required if the pipe annulus contains critical levels of hydrogen sulphide.

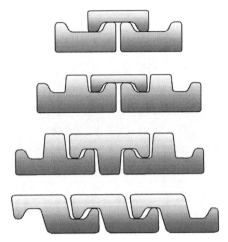

Figure 10.6   Typical pressure armour wires used in unbonded flexible pipes

10.2.5 *Tensile armour*

The tensile armour layer provides the required axial tensile strength capacity to resist pipe weight, end cap load and external tension loads. It is made of helically wound metallic wires with a low lay angle (typically 25°–35°) relative to the longitudinal pipe axis. The wires normally have a rectangular cross-section, but round or profiled wires may be used as well. It is important for the quality of the pipe that the tensile armour wires (except for the round ones) are properly pre-formed during manufacture to obtain the correct helical shape. The wires are cross-wound in pairs to obtain the best possible torsionally balanced pipe design. Therefore, loads due to axial tension and pressure do not induce significant twist or torsional loads in the pipe.

The armour wires are normally made of low-alloyed carbon steel grades with high yield strength, typically in the order of 700–1500 MPa. Steel grades with lower yield strength may, however, be required if the pipe annulus contains critical levels of hydrogen sulphide.

In a product family II design (no dedicated pressure armour) the tensile armour is normally applied with a lay angle close to 55 degrees to balance the radial and axial loads of the pipe. Thereby it is possible to design a pipe which does not elongate when pressurised (both family I and III pipes elongate slightly when pressurised). This circumstance may be utilised to design pipes that are not susceptible to upheaval buckling phenomena. In particular, by specifying a lay angle slightly less than 55 degrees (e.g. around 53–54 degrees) the pipe can be designed to shorten in length when pressurised to compensate for the temperature-induced elongation.

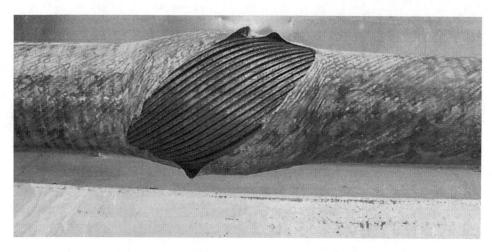

Figure 10.7 Typical 'bird cage' failure mode due to axial compression load

10.2.6 *Holding bandage*

A special high strength bandage may be applied around the tensile armour layer to control the radial displacement of the tensile armour wires. This is particularly important for pipes operating in deep water, as the external hydrostatic pressure can induce high axial compression load in the pipes (reverse end cap effect). The magnitude of the compression load is governed by the hydrostatic pressure differential between the pipe bore and the ambient seawater. As a result, the tensile armour wires may relieve their compression stresses through radial buckling, thus rupturing the outer sheath. An example of such 'bird cage' failure mode is shown in Figure 10.7.

The bandage, which typically consists of tapes made of a fibre-reinforced polymer material, is wound around the outer tensile armour (alternatively also around the inner tensile armour) with a high lay angle. However, restraining the tensile armour in the radial direction may initiate a higher-order lateral buckling mode in the tensile armour (see also 'Pressure and tensile armour' in Section 10.3.3).

10.2.7 *Outer sheath*

The polymeric outer sheath prevents the steel wires in the pipe from being in direct contact with seawater, and provides mechanical protection to the outer tensile armour layer. It is normally made from the extrusion of specific grades of polyethylene (e.g. MDPE) or polyamide (e.g. PA11) materials. PA11 is typically used in dynamic risers because of its relatively good mechanical strength, whereas MDPE is mainly used for static applications, such as jumpers and flowlines.

10.2.8 *Additional layers*

Additional polymeric layers may be required to achieve specific properties of the flexible pipe, for example:

- anti-wear layers;
- intermediate sheath layer;
- thermal insulation layers.

The purpose of the anti-wear layers is to separate the metallic armour layers, and to minimise the friction, so significantly improving the fatigue performance of a dynamic riser. The anti-wear layers are typically made from polymeric tapes (e.g. PA6, PA11) with a thickness of 1–3 mm. Flexible pipes in static applications do not require anti-wear tape.

An extruded intermediate sheath may be applied between the pressure and tensile armour layers. Thereby, the collapse capacity against external hydrostatic pressure loads can be improved by using the collapse capacity of the pressure armour layer. In particular, this option may be adopted for smooth bore pipes, e.g. to enable the pipe to be installed in an empty condition. However, the pressure build-up in the annulus between the inner liner and intermediate sheath should be controlled during the design life. Otherwise, this design principle can be associated with a critical failure mode as excess pressure in the annulus, for example, due to permeation of bore fluids or a leaking inner liner, will be taken by the tensile armour. Thus the combined action from tensile and pressure loads may ultimately result in rupture failure of the tensile armour, for example at the top of a riser.

Flexible pipes have inherently good thermal insulation properties compared with rigid steel pipes owing to their polymeric layers and cross-sectional build-up. However, further passive insulation may be required, such as for long flowlines or risers in deepwater applications. Typically, polymeric tapes consisting of a thermal insulating material (e.g. polypropylene) are wound around the tensile armour layer; see Figure 10.8. Alternatively, an intermediate sheath may be placed between the tensile armour and the insulation. Due attention should be given to the creep resistance of the insulation material, for example, to resist the squeeze loads from clamps.

10.3 Flexible pipe design

10.3.1 *General*

The prediction of the mechanical behaviour of a flexible pipe depends on the interaction between its individual component layers. As this interaction is not straightforward to formulate, a well-documented design procedure is required, possibly supplemented by third party verification/certification.

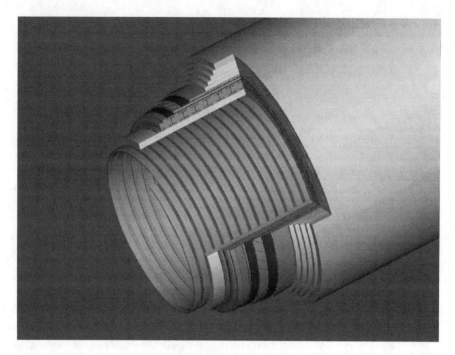

Figure 10.8   Insulated flexible pipe

In recent years several computer programs have been developed for the static and dynamic analyses of flexible pipe systems. To obtain confidence in the analysis results it is important that representative design cases be established, against which the computer programs can be verified. Thus the following points should generally be considered when qualifying a design methodology for flexible pipes:

- The pipe model should be based on a realistic theoretical description.
- Representative design cases should be used to verify the computer program.
- The results should be checked using independent computer programs.
- A verification of the technical requirements should be carried out using proto-type tests.

The design of a flexible pipe system is an interactive process, involving static cross-sectional pipe design combined with system configuration analyses. The static cross-sectional analysis determines the strength capacities against internal pressure, temperature, fluid composition and external loads. This design is then used as input to the system configuration analysis, where the pipe is analysed under all functional, environmental and accidental load combinations. The cross-sectional analyses are normally carried out using the propriety software of the flexible pipe manufacturer.

10.3.2 *Load conditions*

Loads are conveniently classified as functional, environmental, installation and accidental. The flexible pipe should be analysed for the load conditions for:

- factory acceptance testing;
- storage and transportation;
- installation and commissioning;
- normal operation (recurrent and extreme);
- abnormal operation.

Accidental loads should be considered in connection with extreme and abnormal operation. The accidental loads may comprise dropped objects, trawl board impact, failure of vessel positioning system (e.g. thruster and/or anchor line failure), failure of turret drive system, unintended flooding of compartments of subsea arches, and loss of buoyancy modules.

Factory acceptance testing

A hydrostatic pressure test should be carried out as part of the factory acceptance testing to prove the structural integrity of the pipe with end fittings. Normally, this is carried out while the pipe is situated on a reel or turntable.

Storage and transportation

The pipe, including end fittings, accessories and reels should be sufficiently secured and protected during storage and transportation to avoid damage of the pipe and to safeguard personnel. Typically, the loads to consider include:

- loads due to exposure to the environment (e.g. sand particle abrasion, sunlight (UV) ageing);
- acceleration loads during sea transport.

Installation and commissioning

All relevant static and dynamic loads in the flexible pipe during installation and commissioning should be identified, including:

- imposed tension during installation, commissioning and repair;
- loads from the installation equipment;
- tie-in loads including pull-in of risers via J- or I-tubes;
- loads from installation clamps and other supporting/lifting equipment;
- external hydrostatic pressure loads;
- loads from trenching and/or rock dumping operations;
- field hydrostatic pressure testing (as-installed and/or during commissioning);
- loads from inspection tools.

It is of utmost importance that the minimum allowable bend radius for the pipe is not violated during the installation operation. In particular, the lay tension should be sufficiently high to avoid over-bending of the pipe at the touch down point on the seabed.

Loads from the installation equipment on the lay vessel may result in combined crushing and tension loads (e.g. from 2-, 3- or 4-track caterpillars) and combined bending and tension loads (e.g. due to pipe installation via a chute or wheel).

In deepwater installations the external hydrostatic pressure can induce high axial compression load in the pipes (reverse end cap effect), which may result in 'bird cage' failure as described in Section 10.2.6. It is particularly important to assess this failure mode during the installation analysis, as it may be initiated at the seabed touch down area due to the combined action of the 'reverse end cap' compression, dynamic motion induced axial compression, and bending of the pipe.

Normal operation

The annual probability of occurrence for the combined normal operational loads (recurrent and extreme) should be equal to or greater than 10^{-2} for a typical 20-year service life. The functional, environmental and accidental loads to be considered during normal operation comprise:

- vessel first order responses;
- vessel low frequency horizontal excursions;
- response of supporting structure (e.g. submerged arches);
- waves, current and hydrostatic pressure;
- pipe functional loads;
- accidental loads with a probability of occurrence greater than 10^{-2}.

An example of an accidental load under normal operation could be the failure of one mooring line, whereas failure of two mooring lines may be categorised under abnormal operation (due to the statistical difference in their probability of occurrence).

A high internal pressure may significantly reduce the fatigue life performance of a flexible riser due to high inter-layer contact pressure and its effect on the stress variation in the armour wires during flexure. Consequently, if the maximum operational pressure can be established with confidence, this may be used instead of the design pressure when calculating the fatigue life.

Abnormal operation

Possible abnormal load conditions that could occur during the service life should be analysed. The combined annual probability of occurrence for these loads may be considered to be in the range 10^{-2} to 10^{-4}. Load combinations with a yearly probability of occurrence of less than 10^{-4} may be disregarded.

Table 10.2   Design MBR requirements (API Spec 17J)

MBR	Design criterion
Storage	1.1 times the MBR causing locking in armour wires 7.7% strain for PE and PA 7.0% strain for PVDF
Static applications	1.0 times storage MBR

Dynamic applications
  Normal operation
  Abnormal operation
	
	1.5 times storage MBR or 3.5% strain for PVDF
	1.25 times storage MBR or 3.5% strain for PVDF

10.3.3   Cross-sectional design

The flexible pipe structure should be shown to meet the design requirements for each individual layer in the pipe wall for all relevant load conditions. In this section the main design criteria required to fulfil the mechanical integrity of the pipe structure are described.

An important design criterion is the minimum bend radius (MBR) for the pipe, which is governed by the allowable strain of the polymeric layers and permissible relative movements of the wires in the metallic armour layers during pipe bending. Table 10.2 summarises the design MBR requirements according to API Spec 17J.

The MBR requirement along the pipe during installation is normally 1.5 times storage MBR or 3.5% strain for PVDF. When the pipe is installed via equipment with well-defined curvature geometry (e.g. chute or wheel) the MBR requirement for the equipment may be governed by the crushing capacity of the pipe due to the combined action of tension and bending. For a low installation tension the static MBR requirement may be applicable for the installation equipment.

Carcass
The internal carcass should be designed against excessive ovalisation, collapse, yielding, abrasion and erosion. The supporting effect of the pressure armour layer may be taken into account when calculating the ovalisation and collapse capacities. The collapse failure mode may be elastic, plastic or combined elastic–plastic, depending on the carcass structure.

A maximum permanent ovalisation of 0.2% is normally considered acceptable based upon the equation:

$$Ovalisation = (D_{max} - D_{min}) / (D_{max} + D_{min})$$

where D is the carcass diameter. Note that this definition differs from the ovality, as specified by the DNV OS-F101; see Section 6.2.6.

Table 10.3 Allowable hydrostatic collapse utilisation for carcass according to API Spec 17J

Water depth, D_{max}	Allowable utilisation
$D_{max} \leq 300$ m	0.67
300 m $< D_{max} < 900$ m	$((D_{max} - 300)/600)\ 0.18 + 0.67$
$D_{max} \geq 900$ m	0.85

The allowable utilisation against hydrostatic collapse failure modes as specified in API Spec 17J is shown in Table 10.3.

The allowable utilisation against yielding is specified in Table 10.4 below.

Inner liner

The required thickness of the inner liner is to a large extent governed by the creep characteristics of the polymeric material, and consequently it depends upon the pressure rating and fluid temperature. When establishing the minimum required design wall thickness due consideration should be given to thinning of the wall when being bent to MBR, shrinkage (e.g. due to loss of plasticiser) and manufacturing tolerances. The maximum allowable reduction in wall thickness below the minimum required design value due to creep into the gaps of the supporting structural layer should not exceed 30% for all load conditions (API Spec 17J). Additionally, the bending strain limitations as specified in Table 10.2 for PE, PA and PVDF should not be exceeded during flexure of the pipe. For other polymeric materials the allowable strain needs to be documented.

There may be other issues determining the design wall thickness, such as requirements to thermal insulation or control of permeation of the bore fluid constituents into the pipe annulus. Also, for flexible pipes without a carcass (i.e. smooth bore pipes) the inner liner wall thickness may be governed by the need to achieve sufficient hydrostatic collapse capacity, or simply to suit the manufacturing process, for example, to cope with the storage radius on reels (prior to applying the armour layers) without suffering local buckling failure.

Pressure and tensile armour

The pressure and tensile armour layers should be designed against yielding, wear and fatigue failure modes. The utilisation against yielding is defined by:

$$Utilisation = \frac{calculated\ stress}{structural\ capacity}$$

The structural capacity may be either the minimum yield strength or 0.9 times the minimum tensile strength. The allowable stress utilisation according to API Spec 17J is given in Table 10.4. Normal practice today is to assume an average

Table 10.4 Allowable stress utilisation according to API Spec 17J

	Service conditions			Installation		Hydrostatic pressure test – factory and field acceptance
	Normal Operation		Abnormal operation			
	Recurrent operation	Extreme operation				
	Functional and environmental	Functional, environmental and accidental	Functional, environmental and accidental	Functional and environmental	Functional, environmental and accidental	
Carcass and pressure armour	0.55	0.85	0.85	0.67	0.85	0.91
Tensile armour	0.67	0.85	0.85	0.67	0.85	0.91

stress for the wires, which is calculated by distributing the total layer load uniformly over the wires in the layer.

Axial compression load in the pipe cross-section may subject the tensile armour wires to a radial displacement (lift), which ultimately could result in a 'bird cage' failure mode as described in Section 10.2.6. Thus the radial displacement of the wires should be limited to a value that does not violate the integrity of the outer sheath, for example, by applying a holding bandage around the tensile armour. Note, however, that restriction of the tensile armour in the radial direction may initiate a higher order, lateral buckling failure mode of the tensile armour wires. Typically, a maximum radial displacement of half the thickness of the tensile armour wires may be allowed.

Outer sheath

The thickness of the outer sheath should be sufficient to maintain its leak proof integrity during installation and operation of the pipe. Special attention should be given to sections of the pipe subjected to abrasion loads, for example those at the touch down point on the seabed or in the bellmouth area of I- and J-tubes. Additionally, the bending strain limitations as specified for PE and PA in Table 10.2 should not be exceeded during flexure of the pipe. For other polymer materials the allowable strain needs to be documented.

10.3.4 *End fitting*

The end fitting should be designed to ensure:

- reliable termination of all the pipe layers;
- leak tightness against internal and external fluid pressure;

- sufficient fatigue endurance life (e.g. ten times the service life);
- reliable transfer of loads to supporting structure (e.g. a riser hang-off arrangement);
- reliable support of any ancillary equipment (e.g. attached bend limiter).

The stress utilisation against yielding for primary structural components in the end fitting should not be higher than those specified for the pipe itself. For pressure-containing parts of the end fitting, API Spec 17J specifies the following design requirements:

Tensile hoop stress $\leq n \times$ *Structural capacity*
Equivalent stress (von Mises or Tresca) $\leq n \times$ *Structural capacity*

where n is the permissible utilisation factor in Table 10.4, corresponding to carcass and pressure armour.

10.3.5 *System design*

System design may be categorised as global static and global dynamic analyses of the flexible pipe(s) under the functional, environment and accidental loads. The global static analyses may comprise:

- on-bottom stability;
- lateral buckling;
- upheaval buckling;
- forced pipe displacements (e.g. those due to thermal expansion of connecting steel flowline);
- flow assurance (including cool-down analyses);
- static riser configuration with and without any vessel/buoy/arch offsets.

A global dynamic analysis should be performed of the entire riser system (except for static risers) by combining the static loads with dynamic environmental loads taking into account:

- movement of riser top;
- movement of supporting structure (e.g. subsea buoys);
- waves and current;
- hydrodynamics of ancillary equipment (e.g. buoyancy modules);
- marine growth.

Sensitivity analyses should be carried out to evaluate the effect of variations in critical parameters, such as density of pipe fluids, slug flow in the risers, wave periods, wave and current direction, buoyancy (and partial loss hereof), seabed conditions, etc. Furthermore, the susceptibility of interference/clashing between risers and with other system components (mooring lines, vessel pontoons) should be evaluated. While contact between risers and mooring lines should always be avoided, it may be acceptable to have contact between risers of the same size and orientation, provided the number of occurrences and the impact loads are small. Typical results required from the dynamic analyses may comprise:

- riser tension (true and effective) and angles at top and base;
- tension (true and effective) and curvature distribution along the riser;
- tension at connecting pipeline or anchor points on seabed;
- riser departure angles at mid-water arches;
- tension in arch/buoy tethers;
- arch/buoy movements;
- clearance between adjacent risers (including attached ancillary equipment);
- clearance between riser system and seabed (e.g. sag bend areas);
- clearance between riser and other system components (e.g. mooring lines).

Whereas vortex-induced vibration (VIV) is critical for the fatigue performance of rigid steel pipes, it is generally not considered a problem for unbonded flexible pipes, owing to their inherent structural damping characteristics. However, high axial tension loads, for example in deepwater risers may reduce their structural damping ability, and consequently the effects of VIV may be more significant in such cases.

Riser configuration

A variety of riser configurations may be applied depending upon the application; see Figure 10.9. The different configurations all function as mechanical devices for damping the dynamic movements at the seabed touch down point. In deepwater applications a free hanging configuration can normally be used. The Chinese Lantern riser configuration requires that the topside heave movements are small.

Furthermore, flexible pipes may be used as dynamic jumpers suspended between two floating structures (or between a fixed and a floating structure) in, for example, a U-shape catenary or a Lazy-W configuration (see Figure 10.1). This concept may be used to export stabilised oil from a large capacity floating production unit (e.g. FPSO or TLP) to an offloading facility, such as a buoy system with a shuttle tanker. Also, dynamic jumpers may be used in connection with a free-standing hybrid tower arrangement in deepwater applications. This concept was introduced in 2001 off the shore of West Africa at 1500 m water depth (the Girassol Project). A bundle (tower) of vertical, insulated steel risers are tensioned via a buoyancy tank at the riser top, which typically is located 50 m below the water line. The flexible jumpers thus connect the top of each riser at the tower with a floating facility, for example FPSO. Instead of a bundled tower, a single steel riser may be used to form a Single Line Offset Riser (SLOR).

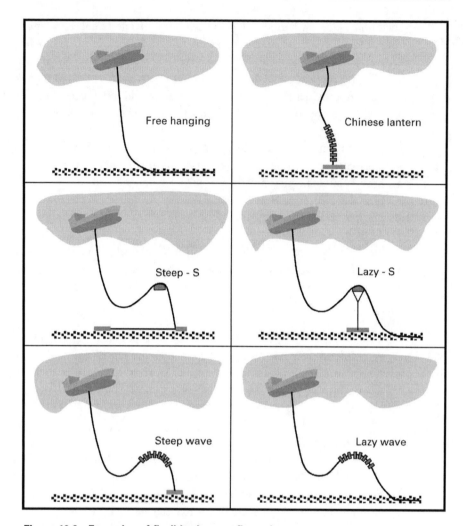

Figure 10.9   Examples of flexible riser configurations

A number of ancillary components are required in a flexible riser system to make it function properly. In particular, it is important to ensure that the MBR design requirement is not violated, and that the global riser configuration does not lose its integrity under extreme load conditions. Dependent upon the system configuration these ancillary components may include:

- bend stiffener;
- bellmouth;
- bend restrictor (static applications);
- subsea buoy;
- buoyancy and weight modules;
- clamping devices;

Figure 10.10   The load out of two subsea buoy systems. Note the reels on the quayside with the risers for the subsea buoys

- riser and tether base;
- tethers;
- abrasion/impact protection;
- riser hang-off structure.

The design loads of the ancillary components are based upon the results of the global static and dynamic analyses. As an example, the design of a bend stiffener requires that simultaneous tension and angle variations be established at the location of the bend stiffener.

A subsea buoy (also called a mid-water arch) consists of buoyancy tanks, which typically are coupled by yokes and clamp support structures. The buoyancy tanks carry an arch with guide gutters and clamps to accommodate the riser pipes. The buoy structure is kept submerged via tethers to a seabed base. Due attention should be given to possible loss of buoyancy during operation, for example that caused by creep of polymeric or composite buoyancy tanks due to the hydrostatic pressure, or by flooding of buoyancy tanks. The buoyancy tanks may be divided into a number of compartments to avoid imbalance in tether tension if one compartment should be flooded. Figure 10.10 shows the load out of two subsea buoy systems with tether bases, each designed for a net buoyancy of 85 t.

Gas venting system
Owing to permeation of fluid constituents through the inner liner, pressure will build up within the pipe annulus, which ultimately may subject the outer sheath to burst failure. Therefore, a gas relieve system is to be designed for controlling the annulus pressure. The system normally forms an integral part of the end fitting, where a number of vent ports (typically three) are evenly distributed

around its circumference. For subsea end fittings check valves are installed in the vent ports, which relieve the annulus pressure at a differential pressure of typically 2–3 bars. For topside end fittings the vent ports are normally connected with a piping system to convey the gas to a safe vent area. It is important that the backpressure of the piping system be sufficiently low to avoid excessive pressure build-up in the riser annulus. When designing the gas relieve system due consideration should be given to:

- gas volume to be vented;
- corrosive attack of the vented gas constituents (e.g. sour service);
- durability in seawater environment (e.g. influence of marine growth).

Fire protection
Fire protection may be applied on a flexible pipe by wrapping or spraying a heat resistant and insulating material on to the pipe body and end fitting. Also, special fire protection caps can be applied on the topside end fitting of risers. Flexible pipes exposed to fire should be considered unfit for further service until detailed examination can demonstrate otherwise.

Pigging and TFL operations
Special requirements may be specified with respect to tools for pigging, TFL (Through Flowline) operations (e.g. according to API RP 17C), work over, or other tool requirements. Special consideration should be given to the pipe design concerning:

- dimensional tolerances (ovality);
- inner diameter variations (e.g. between end fitting and pipe body);
- material compatibility (to carcass).

10.3.6 *Service life analysis*

The prediction of the service life for a flexible pipe requires that due consideration is given to the performance of the applied metallic and polymeric materials with respect to:

- chemical degradation (corrosion of metals and ageing of polymers);
- mechanical degradation (wear, fretting and abrasion);
- fatigue properties (sweet service, sour service);
- time dependent effects (stress relaxation and polymer creep).

Corrosion
The following metallic components should be analysed for possible corrosion attack during installation and operation of the flexible pipe:

- internal carcass;
- steel armour wires in the pipe wall;
- internal wetted surfaces of the end fitting;
- external surfaces of the end fitting and ancillary components;
- components of dissimilar materials (galvanic corrosion).

Special attention should be given to pipes transporting fluids containing carbon dioxide and/or hydrogen sulphide. These constituents can easily permeate through the polymeric inner liner, and thus also subject the armour wires to corrosion attack if liquid water is present in the pipe annulus. A corrosion allowance may be specified for the armour wires and end fitting parts where general corrosion is found acceptable. Furthermore, chloride and mercury constituents in the bore fluids may be critical for certain stainless steel carcass materials, in particular at elevated temperatures.

Preventive measures to mitigate corrosion may include:

- injection of inhibitors in the bore fluid, e.g. to reduce oxygen and bacteria concentrations;
- a corrosion resistant overlay on the internal wetted surfaces of the end fitting;
- electrical isolation of dissimilar steel materials to prevent galvanic corrosion, e.g. between the carcass and the end fitting;
- anti-corrosion coatings;
- cathodic protection (by sacrificial anodes or impressed current).

Submerged end fittings and ancillary equipment are normally protected against corrosion by a suitable coating combined with cathodic protection. The polymeric outer sheath of a flexible pipe provides good protection against seawater corrosion of the steel armour wires, and further corrosion protection of these is normally not required. However, cathodic protection of the armour wires may be specified in some cases to account for possible damage of the outer sheath, for example, by connecting bracelet anodes to sub-sea end fittings. This requires that electrical continuity be present between the end fitting and the steel armour wires. The effectiveness of protecting all the armour wires in a flexible pipe via anodes is not yet fully understood (in particular electrical shielding effects).

Ageing
Polymeric materials are susceptible to ageing when subjected to particular environmental conditions. This degradation process normally affects the material properties unfavourably; in particular, it may reduce the strength and ductility. The ageing mechanism may be associated with a change of the molecular structure of the polymer material, as well as extraction of any plasticiser constituents. Examples of typical ageing environments for flexible pipes comprise:

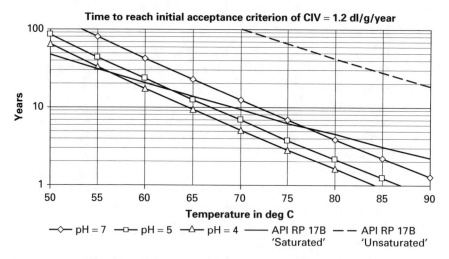

Figure 10.11   Arrhenius plot of service life against temperature for PA-11

- ultraviolet radiation;
- organic acids;
- aromatic fluid constituents;
- injection chemicals;
- methanol;
- hot water (hydrolysis of polyamide).

As the ageing process is strongly accelerated at elevated temperature, lifetime prediction data are normally presented in an Arrhenius plot, defined by:

$$T_{crit} = \alpha \, e^{(-\beta/\theta)}$$

where

T_{crit} critical exposure time at a given temperature θ
α, β constants.

Polyamide materials are susceptible to ageing when exposed to water or methanol at elevated temperature. This hydrolysis phenomenon can be critical for flexible pipes as it may reduce the service life considerably, depending upon the fluid temperature as illustrated in the Arrhenius plot of Figure 10.11 for polyamide PA-11 (API Technical Report TR17 RUG, 2002).

Hydrolysis is a reaction where the polyamide molecule links within the polymer break, resulting in a gradual embrittlement of the material. The rate of ageing depends upon a number of operational parameters, such as water content of the hydrocarbon fluid, temperature, pH-value, organic acids, injection chemicals and fluid composition. The degradation of the polyamide may be expressed as a reduction of its mechanical properties, for example, strain to break, or molecular weight. The change in molecular weight may be measured via its corrected inherent viscosity (CIV).

Sour service

According to API Spec 17J, sour service in a flexible pipe is defined by service conditions with an H_2S content exceeding the minimum specified by NACE MR 01-75 at the design pressure, equal to a partial pressure of 3.5 mbar (0.05 psi).

When assessing the sour service environment of the armour wires in the pipe annulus a simple, and normally conservative, approach is to assume pipe bore conditions. However, a more realistic method is to determine the pipe annulus environment taking into account the permeation characteristics of, in particular, H_2S, CO_2, CH_4 and H_2O. Please note that no corrosion of the armour wires will take place if the annulus is dry. The CO_2 content tends to lower the pH value in the pipe annulus, which increases the susceptibility to hydrogen sulphide induced failure of the wire. CH_4 has the advantage of diluting the annulus composition, reducing the partial pressures of H_2S and CO_2. The permeation model should account for the annulus venting taking place at the vent ports of the end fittings. This point is particularly beneficial for risers having their annulus vented to the atmosphere at the topside end fitting, as it keeps the partial pressure of the constituents at relatively low levels. The effect of possible separation of the annulus fluid should be evaluated, for example gas being trapped at riser hog bend areas or water being trapped at riser sag bend areas. Furthermore, the model should account for permeation through the outer sheath.

The mechanisms involved with respect to cracking, for example sulphide stress cracking (SSC), hydrogen induced cracking (HIC) and stress oriented hydrogen induced cracking (SOHIC) in flexible pipes are extremely complex and not yet fully understood. Furthermore, it is important to note that the standard SSC and HIC tests were developed for the conditions of rigid steel pipes, which are totally different from the special environment of a flexible pipe annulus. In particular, the large surface area of the metallic wires and the limited fluid volume in the pipe annulus are not accounted for. For example, the accumulation of corrosion products (e.g. iron carbonate and iron sulphide scale), the presence of controlling buffer systems (e.g. the carbon dioxide versus bicarbonate buffer CO_2/HCO_3^-) and saturation with Fe^{++} in the annulus will tend to increase the pH value, thus reducing the susceptibility to hydrogen sulphide induced failure. Consequently, a special test protocol reflecting the actual annulus conditions may be used instead of the standard sour service tests. NORSOK Standard M-506 can be used as guidance for the calculation of pH values, based upon the concentration of bicarbonate.

Fatigue analysis

The main contribution to accumulated fatigue of flexible pipes is from the dynamic movements of risers. In particular, the metallic tensile and pressure armour layers are subjected to variations in stress levels, which normally govern the service life duration of risers. Typically, a safety factor of ten is applied to the calculated accumulated fatigue damage. However, a smaller safety factor may be used, depending upon the application and the consequences of failure. Thus a safety factor of five (normal safety class) or even three (low safety class) may be considered. This is analogous to the philosophy adopted in DNV OS-F101 for rigid steel pipes. It should be noted that flexible pipes are redundant structures, owing to their multi-layer construction, and consequently fatigue failure of armour wires would not necessarily result in a catastrophic pipe failure.

The fatigue assessment should be based upon realistic test data of the armour wires. In particular, the fatigue properties of the wires (including welds) should reflect the corrosion effects caused by the pipe annulus environment (e.g. sweet or sour annulus conditions). The fatigue properties are typically presented in design S–N curves, showing number of cycles to failure as a function of stress amplitudes.

Calculations of the accumulated fatigue damage are normally based upon Palmgren–Miner's rule, unless the stress amplitudes in the wires can be shown to be below the endurance limit of the S–N curves. However, such an endurance limit may not exist in the case of corrosion fatigue. The fatigue calculation model should be able to analyse the stick–slip behaviour of the armour wires during pipe bending. Furthermore, the establishment of correct contact pressure and frictional condition for the wires is important to obtain realistic stress variations.

A critical corrosion fatigue case may arise if the annulus of a dynamic riser is flooded with seawater and the seawater is continuously exchanged, for example due to a damaged outer sheath. In particular, the oxygen content of the seawater will subject the armour wires to corrosion fatigue, which ultimately could result in fatigue failure within a short time period, of the order of one year or less.

Condition monitoring

Various methods may be employed to assess the condition of a flexible pipe, including:

- visual inspection;
- test coupons;
- test spool pieces;
- annulus gas monitoring;
- eddy current inspection;
- radiographic inspection;
- optical fibre sensors.

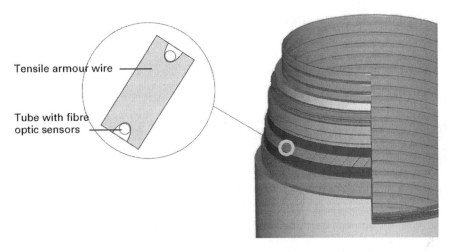

Figure 10.12   Condition monitoring using optical fibre technology

Implementation of optical fibre sensors in the armour wires has been shown to be a viable option (Andersen *et al.* 2001) to obtain the full strain and temperature history of a riser, for example, during operation in the field. Thereby, the actual accumulated fatigue damage can be continuously updated and compared with the design assumption. The sensors may also be used to monitor the environmental condition in the pipe annulus, to assess corrosion, for example. An example of implementation of an optical fibre in a flexible pipe is shown in Figure 10.12.

10.4   Material qualification

10.4.1   *General*

The physical, mechanical, chemical and performance characteristics of the materials used in flexible pipes (including end fittings) should be subject to qualification testing. Typically, the materials currently used in flexible pipes comprise:

- metallic materials;
- polymeric materials;
- epoxy resin;
- fibre-reinforced materials.

As a minimum, the metallic materials should be supplied with a 3.1B certificate (ISO/DIS 10474 3.1B), indicating that the material is traceable, but that no third party certification has taken place. The qualification requirements of metallic and polymeric materials are briefly discussed below.

10.4.2 *Metallic materials*

Metallic materials (e.g. carbon steel, stainless steel, duplex steel) are primarily used in un-bonded flexible pipes, owing to their good mechanical strength, chemical resistance and erosion properties. In the selection of material type and grade, due consideration should be given to the resistance against corrosion (in particular pitting corrosion), sulphide stress cracking (SSC) and hydrogen induced cracking (HIC). For wire/strip materials and weldments the following material properties should normally be documented:

- chemical composition;
- yield strength/elongation;
- ultimate strength/elongation;
- hardness (sour service applications);
- SSC and HIC (if applicable);
- corrosion resistance;
- erosion resistance (carcass only);
- fatigue resistance (e.g. corrosion fatigue);
- hydrogen embrittlement (e.g. cathodic protection of armour wires);
- chemical resistance.

As the carcass material is in direct contact with the bore fluids at operational conditions, its chemical compatibility should be verified in a test environment simulating these conditions. The corrosion testing may be carried out in dedicated autoclaves under representative pressure and temperature conditions, and subjected to the critical fluid constituents, e.g. chloride, carbon dioxide and/or hydrogen sulphide. The testing should be performed on carcass test samples resembling the manufacturing process, such as cold forming and welds, for example.

End-fitting metallic components for primary pressure containment should be wrought or forged. Testing of the mechanical properties of forgings may be performed on integrated test coupons that are removed from the forgings after the final heat treatment. However, it is recognised that pipe standards allow separate test coupons to be heat treated simultaneously with the material they represent. In these cases, special attention must be given to whether the material thickness, forging reduction and mass of the test specimen are representative of the actual forging.

For the end fitting, the following material properties should be documented:

- chemical composition;
- tensile properties;
- hardness (sour service applications);
- Charpy V impact for primary end-fitting components;
- resistance to SSC and HIC (if applicable);

- corrosion resistance;
- erosion resistance.

10.4.3 *Polymeric materials*

The polymeric materials used in flexible pipes serve several functions, from the sealing of the pipe wall (extruded inner liner and outer sheath) to providing thermal insulation, or serving as anti-wear layers. There are in particular three generic semi-crystalline polymers suitable for the large-scale extrusion process required for flexible pipes. These are polyethylene (MDPE, HDPE, XLPE), polyamide (e.g. PA11 and PA12) and polyvinylidene fluoride (PVDF) for high temperature applications. The key requirements of the polymers may be categorised into:

- high long term allowable static and dynamic strains;
- long term internal and external fluid compatibility properties;
- low fluid permeability;
- low susceptibility to swelling, blistering and stress corrosion cracking;
- good mechanical resistance against impact, wear and abrasion.

Qualification testing of the polymeric materials may be performed in accordance with API Spec 17J. It is important that the qualification tests of the inner liner and outer sheath are carried out on specimens from extruded material of the fabrication facility for verification of the actual processing parameters.

10.4.4 *Epoxy resin*

Epoxy is used as a filler material in the end-fitting, particularly in connection with embedding the tensile armour wires to secure against slippage. To obtain realistic tests results, the qualification test specimens should be moulded and cured under identical temperature and humidity conditions as when filling the end fitting. The qualification test requirements for the cured epoxy include:

- compressive strength;
- glass transition temperature;
- fluid compatibility;
- ageing;
- degree of cure.

10.4.5 *Fibre-reinforced polymers*

Owing to advances in technology within fibre-reinforced polymer (FRP) composites these materials have become an important alternative to steel armour in flexible pipes, not only for ultra-deep water applications, but also for applications involving extremely sour service. FRP offers a number of important benefits over traditional steel materials, including high strength to weight ratio, excellent resistance to corrosion and good fatigue behaviour.

The mechanical and chemical characteristics of FRP are generally very different from metallic materials, and as a consequence a simple substitution of metals in flexible pipes is not viable. In particular, one major difference between FRP and metallic materials is that the composite fibres are not suitable for absorbing large compressive loads over extended periods of time. This is due to possible visco-elastic failure of the matrix material, which may ultimately lead to fibre kinking and subsequent failure of the entire composite structure.

No specification is currently available concerning qualification of FRP materials for flexible pipes.

10.5   Fabrication

10.5.1   *General*

The manufacturing of flexible pipes should be carried out according to approved QA/QC procedures and written fabrication specifications. As an example, consideration should be given to:

- raw material identification, handling and testing;
- handling of intermediate and finished pipe products;
- moisture control;
- process control (polymer extrusion, winding of wires and tapes);
- welding, heat treatment and coating;
- end-fitting installation (including bend stiffener);
- defect repair (pipe, end-fitting);
- non-destructive examination;
- testing procedures.

A central document during fabrication is the Manufacturing Quality Plan, specifying all the control procedures, inspection points, test procedures and acceptance criteria that are required to ensure that the pipe product complies with the design requirements.

10.5.2   *Pipe*

The factory specifications form the basis for the manufacture of the flexible pipe, typically specifying for each process:

- fabrication method (including the use of any fabrication aids, e.g. lubricants and tapes);

- applicable work instructions;
- raw material identification;
- welding and heat treatment procedures;
- extent of NDE and acceptance criteria;
- repair procedures (note that repair of the inner liner is not allowed);
- marking requirements for the pipe (e.g. length indications and locations for buoyancy modules).

The as-built documentation may include:

- material certificates;
- production quality records (e.g. receiving inspection, dimensions, batch number);
- acceptance test results;
- pipe chart and weld map;
- welder qualification records;
- welding procedures;
- factory acceptance test results;
- non-conformances and issued Concession Requests;
- release certificates;
- certificate of conformity.

10.5.3 *End fitting*

The mounting of end fittings on to a pipe should be carried out according to procedures specifying how each layer of the pipe is terminated. Special attention should be given to:

- inspection of the exposed pipe layers;
- control of temperature during welding and epoxy curing to avoid damaging the polymeric layers;
- surface finish at sealing areas;
- seal energising procedures;
- epoxy filling procedures;
- extent of NDE and acceptance criteria.

10.5.4 *Factory acceptance test*

After the completion of a flexible pipe with end-fittings, it is normally subjected to non-destructive acceptance tests, which may be witnessed by an independent agent. The extent of testing may vary from project to project, but generally the acceptance tests comprise:

- gauge pigging;
- hydrostatic pressure test;
- electrical continuity and resistance test;
- gas venting system test.

Gauge pigging

Gauge pigging is applicable to rough bore pipes (i.e. pipes with an inner carcass), and should be performed prior to the hydrostatic pressure testing. The purpose of gauge pigging is to verify that the internal diameter of the pipe meets the design requirements. The gauge plate may consist of a 5–10 mm thick disc of, say, aluminium with a prescribed diameter. Typically, the gauge diameter should not be greater than 95% of nominal inner pipe diameter.

Hydrostatic pressure test

All pipes should be subjected to a hydrostatic pressure test to verify the integrity of pipe and end fittings. A test pressure corresponding to 1.5 times the design pressure is typically specified, which should be held for at least 24 hours without sign of leakage. API Spec 17J allows a lower test pressure, corresponding to a minimum of 1.3 times the design pressure, which may be used for flexible flowlines and subsea jumpers. The pressurising, testing and de-pressurising are to be performed according to approved test procedure. The air content of the test water should be kept as low as possible, for example by degassing the water prior to flooding the line and by proper control of the water filling process using pigs.

Electrical continuity test

An electrical continuity test may be specified for cathodically protected pipes to verify that there is electrical continuity between the two end fittings via the tensile armour.

Electrical resistance test

For rough bore pipes, it may be required to isolate the carcass electrically from the end fitting to avoid local galvanic corrosion due to the use of materials of different electrical potential (e.g. a carcass of stainless steel against carbon steel elements of the end fitting). A test measuring of the resistance between the end fitting and the carcass may be performed to verify the electrical resistance.

Gas venting test

The vent system for relieving accumulated gas from the pipe annulus should be tested to document that it functions according to specification.

10.5.5   *Packing and load out*

The section length of a flexible pipe is the governing factor for the way it is packed for shipment. Short jumper sections may be delivered in a straight or coiled configuration, secured on dedicated crate or pallet supports. Pipe sections longer than, for example, 40–50 m may be spooled on dedicated offshore reels. The reel can be compartmentalised to accommodate different pipe sections, or simply to separate the end fitting and possible ancillary equipment (e.g. bend stiffener) from the pipe windings. Spooling of a flexible pipe on to a reel or basket should be carried out with a sufficient tension load to avoid the pipe windings from being slack during the subsequent unreeling at the installation location. Furthermore, the end fittings and possible accessories should be properly supported and safely secured to sustain the acceleration loads during sea transport. The pipe may be shielded from the environment using a tarpaulin cloth. The load out of crates, pallets, baskets, reels or ancillary equipment includes lifting the product on-board the transportation/installation vessel, followed by proper sea fastening and protection.

Long pipe sections that cannot be accommodated on reels are normally spooled on turntables. These are fixed installations, and consequently it is required that the pipe be spooled from the onshore turntable to a dedicated turntable on the installation vessel. Typically, a caterpillar is located at the quayside to facilitate the spooling operation.

The pipes may be delivered in empty or flooded condition. Empty pipes that require long time storage may be purged with nitrogen. Pipes that are delivered water filled should be sufficiently inhibited, taking into account the time and temperature dependent degradation of the inhibitor. If the pipe is intended for dry gas service it may be delivered filled, for example, with glycol. Installation aid devices may be part of the scope for the pipe delivery, including pull heads and clamps for lifting/holding the pipe (e.g. Chinese fingers).

10.6 Installation and pipe qualification

10.6.1 *Installation*

Flexible pipe installation is characterised by being a fast operation that can use even a relatively narrow weather window. Furthermore, the bending compliance of the product facilitates the use of a catenary pipe lay configuration with a minimum of back tension. The installation is carried out from dedicated offshore vessels suitable for handling flexible pipes without violating their integrity. The flexible pipes are usually spooled from reels or turntables on the vessel, whilst a sufficient back tension is maintained to avoid over-bending the pipe at seabed touch down. The back tension is normally obtained by the use of one or more caterpillar tensioners with two, three or four tracks. Alternatively, the back tension can be provided via winch driven reels. The flexible pipe may leave the vessel via a wheel or a chute, or directly from a caterpillar in a vertically inclined position (J-lay configuration). Short jumper sections may be installed using lifting equipment.

Trenching of a flexible pipe may be required to protect it from third party damage, for example from fishing equipment, or to obtain additional thermal insulation properties. The trenching operation may be carried out simultaneously with the pipe laying, or subsequent to it.

Installation of a flexible pipe that is susceptible to pressure-induced upheaval buckling may require special attention. A way to mitigate the upheaval buckling problem is to pressurise the pipe during the laying or trenching operation. Thereby the pressure-induced elongation will be converted to an axial pre-tension when the pipe is de-pressurising, due to the frictional soil resistance along the pipe section.

The installation analysis should comprise a verification of loads and displacements, which are critical for the integrity of the pipe. In particular, the following need to be considered for the pipe:

- axial tension capacity;
- crushing capacity;
- slippage of outer sheath;
- acceptable bend radius at vessel (high tension and bending);
- acceptable bend radius at touch down point at seabed (low tension and bending).

Due consideration should be given to the geometry and surface finish of the installation equipment in contact with the flexible pipe. For example, special pad geometries may be used in the caterpillar tensioners to provide a good grip on the outer sheath of the pipe, thus safely transferring the radial contact loads to the tensile armour layers. In particular, the risk of pipe slippage (at low radial loads) or pipe crushing (at high radial loads) should be carefully evaluated. Pipe slippage can occur between the tensioner pad and outer sheath interface, or between the outer sheath and underlying tensile armour interface. The pads may be made of a hard rubber material with a flat or V-shaped geometry (e.g. with a 120°–140° opening angle). The surface finish of chutes should be sufficiently smooth to avoid abrasion/scratching of the polymeric outer sheath. Also, it may be necessary to wet the chute to reduce the friction with the pipe. Soft slings or special handling clamps should be used when lifting/holding the pipe or end fitting, and precaution should be made to avoid over-bending the pipe. In particular the section just behind the end fitting is susceptible to localised bending.

The installation of risers may require specific ancillary equipment to be attached during the installation operation, for example buoyancy/weight modules, abrasion/impact protection and clamp devices. It is important that the clamping contact pressure be specified to account for creep of the polymeric outer sheath/insulation and changes in pipe diameter due to axial tension, pressure and temperature loads.

It is recognised that a majority of flexible pipe failures occur during the installation phase. A common failure is damage of the outer sheath due to interference with, for example, steel wires or trenching equipment. This may result in seawater entering the pipe annulus, subjecting the steel armour to corrosion attack. Consequently, repair of the damaged outer sheath needs to be carried out as soon as possible, for example by the use of a clamp, to prevent the exchange of seawater in the pipe annulus.

To avoid damaging the pipe it is important to manage the installation operation in a safe and controlled manner. This would include documentation of critical parameters such as:

- laying tension;
- departure angle;
- tensioner compression load;
- bending radius at touch down point (e.g. using ROV mounted camera).

10.6.2   *Flexible pipe qualification*

The overall objective of establishing a qualification scheme for flexible pipes is to verify and document quality, reliability and safety of pipe products, thus minimising the possibility of failure during operation of the pipe over its service life. Therefore, all the parties involved in a flexible pipe project, including the pipe manufacturer, owner and insurance company, share a mutual interest in establishing a set of generally acceptable qualification requirements to secure the performance characteristics of the flexible pipes.

To qualify a flexible pipe product fully, due attention should be given to the applied design philosophy, long term performance characteristics of the materials used (in particular polymers), fabrication means and methods, as well as the conduction of relevant prototype tests. However, even though a full size standard test gives a certain confidence of a pipe's capability, it cannot fully reflect the actual conditions under which the pipe will be operated.

An issue of paramount importance in the qualification of flexible pipes concerns QA/QC during design and manufacture. Flexible pipes are very complex due to the composite build-up and the variety of manufacturing processes involved. Therefore, design methods cannot be dissociated from fabrication means and methods, nor can proper QA/QC procedures be neglected.

The majority of qualification tests are destructive, involving full scale testing of a number of pipe sections with end-fittings. Consequently, these tests become extremely resource demanding and costly to perform. Thus the test programme should be planned and carried out in close co-operation with the oil industry to ensure that all the relevant aspects of the pipe performance are taken into account.

10.6.3   *Prototype testing*

Prototype tests are performed to confirm that the technical requirements of a specific flexible pipe product are complied with. These tests are normally verified

Figure 10.13 Flexible pipe during a burst test

by an independent agent. The required amount of testing depends upon the intended service conditions, for example static or dynamic applications, H_2S/CO_2 service, high temperature, etc. The available specifications of API RP 17B, Bureau Veritas and DNV distinguish between standard tests and special tests. Standard prototype tests may be regarded as the minimum recommended type of tests, covering the most common cases of flexible pipe applications. Typically, these tests comprise a burst pressure test, axial tension test and hydrostatic collapse test. Figure 10.13 shows a flexible pipe subjected to a burst test. In addition, special tests may be required for special applications, such as long-term dynamic behaviour, extreme service conditions or installation related items. Table 10.5 presents the classification of prototype tests according to API RP 17B, applicable for unbonded flexible pipes.

No exact requirements exist when assessing the validity range of a prototype tested product, for example limitations on pressure, temperature, pipe dimension and function. As a guidance, a scaling comparison may be applied based upon pressure by internal diameter ($p \times ID$), where the test pipe qualifies pipes with an equal or lower $p \times ID$ value, and assuming that the pipe is based upon the same design methodology, fabrication procedures, and is intended for the same service function as the test pipe. API RP 17B additionally recommends that the above scaling comparison be subject to an internal diameter limitation of 2″ larger or smaller than the test pipe.

The dynamic fatigue test is costly and time consuming to carry out. Typically, a pipe section with end-fittings and bending limiter (bend stiffener or bellmouth) is hung vertically from a rocker arm that imposes cyclic bending to the riser top end; see Figure 10.14. The testing is performed with the riser pressurised and under tension. Alternative test equipment exists where the riser is fatigue tested in a horizontal configuration.

A distinction is made between two types of dynamic fatigue tests: a service simulation test and a service life model validation test. A service simulation test is carried out to determine the structural integrity of the riser and bend limiter while subjected to operational load conditions. The test programme is established

Table 10.5   Classification of prototype tests applicable for unbonded flexible pipes (API RP 17B)

Test type	Description	Test condition/Comment
Standard prototype tests (Class I)	Burst pressure test Axial tension test Hydrostatic collapse test	Typically in straight line At ambient pressure With outer sheath perforated or omitted
Special prototype tests (Class II)	Dynamic fatigue test	Bending, tension, torsional, rotational bending or combined bending and tension fatigue tests
	Crush strength test Combined bending and tensile test	Installation test Installation test
	Sour service test Fire test Erosion test Through flowline (TFL) test	To examine degradation of steel wires To examine degradation of carcass Also includes pigging test
Other prototype tests (Class III)	Bending stiffness test Torsional stiffness test Abrasion test Rapid decompression test Axial compression test	To MBR (non-destructive) To allowable torque (non-destructive) Test for external abrasion Upheaval buckling and compression capacity
	Thermal characteristics test Temperature test Arctic test Weathering test Structural damping test	Dry and flooded conditions High and low temperature cycling Low temperature test UV resistance Characterisation test

on the basis of a dynamic simulation analysis. This is done by selecting the governing fatigue loads, which typically represent approximately two million cycles (out of 100 million cycles during a 20-year design life, for example). The test duration depends upon the obtainable cyclic frequency, but a four to six month test period should be anticipated to achieve two million fatigue cycles. A service life model validation requires that the test be carried out until an accumulated fatigue damage of one is reached, i.e. it is a destructive test. Thereby, it is possible to validate and calibrate the fatigue analysis tools.

Verification test protocols – high temperature application

It is of paramount importance that prototype tests reflect the intended service conditions of the pipe as accurately as possible. However, as described in the following, field experience from 1994 and 1995 revealed that the conduction of the existing standard prototype tests did not always ensure that a particular pipe product was fit for purpose.

Owing to the increasing demand for higher temperature rating for flexible pipes in the early 1990s, special plasticised PVDF material grades were used

Figure 10.14 In-plane bending fatigue test equipment

without reservation, though only limited field experience existed (in particular for dynamic applications). Unfortunately, a number of leak failures were experienced by the oil companies during 1994 and 1995 in oil/gas risers, operating at temperatures above 80°C. These leak failures were predominantly associated with the termination within the end fitting of multi-layer extruded inner liner fabricated from plasticised PVDF homopolymer. Owing to the extraction of the plasticiser during operation of the pipes the inner liner tube was subjected to shrinkage, which eventually resulted in slippage of the inner liner from the sealing area.

Consequently, to assess the susceptibility to end-fitting failure when using plasticised PVDF, a series of owner meetings were held during 1995 and 1996 to attempt to reach an oil industry agreement on qualification protocols for both static and dynamic pipes. This resulted in a new test protocol for the qualification of high temperature end fittings with the objective of verifying the seal integrity. The basic idea behind the new test procedure is that temperature cycling will gradually extract (and partially replace with hydrocarbons) the plasticiser within the polymer of the plasticised grades. Furthermore, the temperature variation will impose alternating axial loads in the inner liner tube, which could be critical for the seal system. The protocol was later extended to cover flexible pipes with inner liner tubes made of non-plasticised material grades (those that have only 2% (by weight) volatile content), and this protocol was subsequently included as an Appendix to API RP 17B.

Chapter 11
Operation

11.1 Introduction

Written procedures should be established for the safe commissioning, operation, inspection, maintenance, repair, modification and abandonment of a pipeline system. The procedures should be subject to revision as required, for example, due to changes in the condition of the pipe system. Furthermore, attention should be given to appropriate training of personnel, possible liaison with third parties and retention of records.

The operation and maintenance procedures should, at a minimum, provide information with respect to:

- organisation and management;
- start-up and shut-down procedures (including emergency shut-down procedures);
- operational limitations;
- identification of all items to be monitored, inspected and maintained;
- specification of monitoring equipment, inspection methods, frequency of inspection and acceptance criteria;
- pigging requirements.

A specific corrosion management system should be established to minimise the risk of pipe failure and loss of operability. This may include regular pigging to prevent the accumulation of corrosive liquids, as well as monitoring of any inhibitor programmes. Inhibitors may be routinely injected to reduce internal corrosion, but also to mitigate water drop-out or the formation of wax, paraffin or hydrates.

Emergency procedures should be prepared to ensure that all personnel and others likely to be involved are adequately informed on the actions to be taken in the event of an incident or emergency.

Separate procedures covering non-routine or special activities should be prepared when required, for example for major repairs, modifications or abandonment of a pipeline system, or part of a pipeline system.

11.2   Flow assurance

11.2.1   *General*

The implementation of a flow assurance strategy is of utmost importance for preventing production losses due to blockage of the pipe transportation system during its service life, for example that caused by the formation of hydrates, wax, asphaltenes or scale deposits. The flow assurance concept may also be extended to include corrosion, erosion by sand and/or droplet impingement, formation of water–oil emulsions with high velocity, slugging and other operationally undesirable hydraulic phenomena.

The flow assurance assessment forms an integral part of an offshore field development, including concept selection, project execution, asset operation and system surveillance. Furthermore, flow assurance needs to be considered for the entire fluid system from reservoir to the receiving facilities. As an example, the fluid condition in a production riser depends upon the fluid processes taking place in the connecting pipeline system.

Slug flow may result in undesirable mechanical loads in a riser system and the receiving process equipment. Also, slug flow may prohibit the process facilities from being operated in an optimal way. These problems become increasingly greater in deep-water developments due to the greater change in pressure and temperature conditions from reservoir to receiving facilities than in more shallow developments.

11.2.2   *Mitigation of flow blockage*

The mitigation of flow blockage may be achieved by adopting various solutions, which may be applied in combination. Table 11.1 gives an overview of typical mitigating measures for preventing flow blockage.

Hydrates in pipeline systems are typically controlled by maintaining the fluid temperature above the hydrate formation temperature. This is not a problem during steady state fluid conditions if the temperature profile along the line can be kept sufficiently high. However, for long flowlines and during start-up and shut-down of a pipeline, the temperature may drop below the critical temperature, which necessitates that additional action be taken to prevent the formation of hydrates. In such cases, insulation of the pipeline in combination with methanol or glycol injection may be applied in order to control the hydrate formation.

A number of chemicals may be applied to mitigate flow blockage, including methanol, hydrate inhibitors, asphaltene inhibitor, paraffin inhibitor, pour-point depressant, scale inhibitor and corrosion inhibitor. However, these chemicals are usually costly as they are not straightforward to handle, they require special storage and delivery systems and their contamination impact on, for example, crude oil and condensate, may reduce the market price of these. Furthermore, they

Table 11.1 Mitigating measures for preventing flow blockage

Mitigating measure	Characteristic means
Thermal insulation	Materials with a low thermal conductivity
Active heating	Electrical heating, hot fluids
Chemical usage	Methanol, glycol (MEG), inhibitors
Pigging	Removal of deposits and liquids
Operational procedures	Control of fluid temperature and pressure
	Control of fluid humidity
	Filling the pipe with diesel

may have unfavourable environmental impacts, which need to be addressed. Consequently, there is an incentive to reduce the use of these chemicals by applying special flow assurance strategies. This may include methods involving active heating of the pipeline, the use of 'cold flow' concepts where solids are allowed to depose in a controlled manner, or the introduction of special 'low dosage' inhibitors.

11.3   Operation, maintenance and abandonment

11.3.1   *General*

A management system should be established and implemented to ensure a safe and efficient operation, maintenance and abandonment of the pipeline system. Furthermore, the management system is an important tool to ensure compliance with the design and to deal effectively with incidents and modifications.

11.3.2   *Commissioning and operation*

Pre-commissioning activities are described in Section 8.8, and typically comprise hydrostatic pressure testing, cleaning of the pipe with its components, gauging and possibly drying in the case of gas pipe systems. These activities may be carried out as part of the installation scope of work. The subsequent commissioning activities prior to operation comprise functional testing of the system and equipment, and activities associated with the initial filling of the pipeline system with the fluid to be transported. In particular, it is important to test the equipment used for monitoring and control of the pipeline system. This includes safety systems associated with pig-trap interlocks, pressure-, temperature- and flow-monitoring systems, and emergency pipeline shutdown systems. Furthermore, it should be checked whether valves are operating satisfactory.

Filling of the pipeline system with the fluid to be transported should be carried out according to written start-up procedures. Prior to introducing the fluid into the system it should be ensured that all functional testing is complete and accepted, and that operational procedures are in place. Also, a formal transfer of the pipeline system to those responsible for its operation should be completed.

It is important that critical product parameters are kept within the specified design limits during operation of the pipeline system. In particular, the design pressure and temperature should not be exceeded during normal operation. The maximum allowable operating pressure (MAOP) should be established with a sufficient safety margin, taking into account the accuracy tolerances of the pressure control devices. Furthermore, surge pressures should not exceed pre-defined values (e.g. 1.05 times the internal design pressure), and it should be documented that the specified valve closure time does not result in transient pressures exceeding the maximum allowable surge pressure.

The following product parameters should at least be considered:

- pressure control and monitoring along the pipeline system;
- temperature control and monitoring along the pipeline system;
- dew point control and monitoring for gas lines;
- product composition, flow rate, density and viscosity;
- parameters as defined in the corrosion management system;
- concentration of toxic product constituents.

The safety equipment in a pipeline system should be tested and inspected at agreed intervals to verify that it can properly perform the safety function and that the integrity of the safety equipment is intact. This includes pressure control and over-pressure protection devices, emergency shut down systems, automatic shut down valves, and other safety equipment.

Check valves are normally kept open by the fluid flow, and will close in the event of an upstream pressure drop. A mechanical override (operated manually or by ROV), allows the valve to be opened, but it cannot be operated against a significant pressure differential. Therefore a check valve is provided with a by-pass for pressure equalisation, which may be a separate (small diameter) bypass line or may be built into the valve. The check valve can be fixed in the open position to allow bi-directional flow and pigging. The operating position normally permits uni-directional pigging, although some pigs may damage the valve if it is not fixed in the open position.

Ball valves can be opened and closed without pressure equalisation, either manually (by diver) or by ROV, or by means of an actuator, which may be remotely controlled. Subsea isolation valves (SSIV) are always remotely operated. Ball valves normally allow bi-directional pigging.

Figure 11.1 shows a typical valve assembly at a riser base, with the check valve at the riser side. This is because in the event of rupture of the riser the check valve will close as a result of the differential pressure, but it can then be isolated by closing the ball valve and opened by the mechanical override. If it was on the downstream side of the ball valve it could not be opened again

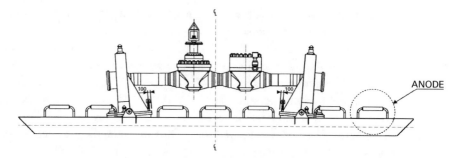

Figure 11.1 Ball valve (left) and check valve (right) on skid-mounted riser base

without de-pressurising the entire pipeline. In shallow water there is also the possibility of replacing the check valve, which has a shorter design life than the ball valve.

11.3.3 *Maintenance*

Maintenance procedures should be prepared to monitor the condition of the pipeline system and to provide the information required to assess its integrity. In particular, the following should be considered when establishing the survey inspection requirements:

- as-built condition;
- results from start-up, periodical and additional surveys;
- predicted integrity condition of the pipeline system;
- required inspection time interval;
- relevant legislation and statutory authorities.

It is important that the as-built data are fully up-to-date for making a proper evaluation of the survey observations, for example, for establishing trend lines and their discrepancy from the prediction. The survey data may be provided via in-service inspections, which are described in more detail in Section 11.4.

11.3.4 *Changes to the design condition*

It should be demonstrated that changes in the design conditions of a pipeline system do not violate its integrity, and that the pipeline system still meets the requirements of relevant standards and legislation. Any changes to a pipeline system should be fully documented prior to implementation. Examples of changes to a pipeline system include:

- change in environmental data;
- change of functional data (e.g. pressure, temperature, fluid type);

- extension of design life;
- modifications due to new developments in the vicinity;
- damage to pipeline system.

Changes in a pipeline system may require that additional hydrostatic testing and inspection be carried out (re-qualification).

11.3.5 *Decommissioning and abandonment*

A pipeline that will be out of service for an extended period may be decommissioned. However, the pipeline should still be properly maintained and cathodically protected. The decommissioning work should comprise removal of service fluids from the pipe bore and isolation of the pipeline section from other parts of the pipeline system left in service. Liquids may be pumped, or pigged, out of the pipeline using water or an inert gas, considering:

- buoyancy effects if gas is used to displace liquids;
- compression effects which may result in ignition of fluid vapour;
- possible asphyxiating effects of the used inert gases;
- disposal of the pipeline fluids;
- drainage of 'dead sections', e.g. valve cavities;
- combustibility of the displaced fluids.

If a pipeline is to be abandoned it should be decommissioned, disconnected from other installations and left in a safe condition. Local legislation may require that the abandoned pipeline be finally removed.

11.3.6 *Pipeline pigging*

Pigging of pipelines is mainly performed to:

- separate dissimilar fluid products;
- displace debris, fluids or deposits (e.g. cleaning, dewatering, removal of wax and scale);
- inspect the pipe.

The pigging operation is carried out using purpose-built pigs that are inserted into the pipeline via a pig launcher and driven by the product flow or special drive fluid (e.g. water, glycol, methanol, diesel). The pigs are recovered again at special receiving stations, so-called pig traps. There are a variety of pigs intended for cleaning and sealing purposes. These include foam pigs, mandrel pigs, solid cast pigs and spheres, which all have the ability to seal against the pipe inner surface, and thus may be used to separate dissimilar products or to sweep liquids from the pipeline. Special gel pigs have been developed, which are highly viscous liquid batches pumped or driven through the pipeline by conventional seal pigs; see Section 8.8.4. Most pipeline gels are water-based, but also chemicals, solvents, acids or diesel may be gelled. The choice of specific gel depends upon the pigging purpose, for example, product separation (batching), debris-, condensate- and water removal, or provision of corrosion inhibitors (gas pipelines). Special intelligent pigs may be used to inspect the pipeline bore, including pipe geometry, corrosion and crack detection, pipeline profile/curvature, leak detection and measurement of wax deposition.

11.4 In-service inspection

11.4.1 *General*

Inspection surveys should be carried out at regular intervals to verify the integrity of the pipeline system. Where special events have taken place that may impair the safety, reliability, strength or stability of the pipeline system, additional inspection surveys should be carried out. When anomalies are found during an inspection survey remedial actions may need to be taken. In such cases a strategy for re-establishing the integrity of the pipeline system should be worked out.

Normally inaccessible parts of the pipeline system, such as a riser pipe located within a caisson, may be remotely examined using suitable equipment, intelligent pigs, for example. The riser condition can then be assessed on the basis of combined internal video inspection (e.g. corrosion type), eddy current examination (e.g. cracks, pitting holes) and ultrasonic measurement (e.g. pipe wall thickness).

11.4.2 *Start-up inspection*

The objective of the start-up inspection survey is to verify that the system is acceptable for service, and to observe any movements or abnormal behaviour of the pipeline system during and immediately after the pipeline system is put into operation. In addition, it provides a baseline for subsequent inspection surveys. The following may be considered:

- verification that the required overburden is present along pipelines (e.g. pipes that are susceptible to upheaval buckling);
- inspection and measurement of pipeline expansion and riser displacements caused by pressure and thermal effects;
- visual inspection of mechanical couplings and flange connections;
- leak detection;
- verification that the operational parameters are within acceptable limits;
- verification that instrumentation is properly installed and is functioning.

Furthermore, it is important that the corrosion monitoring and recording equipment has been installed as agreed (e.g. position of reference points for ultrasonic wall thickness measurements, and location and position of corrosion loss coupons), and that basis measurements have been established.

11.4.3 *Periodical inspection*

Periodical inspection surveys should be carried out at specified intervals, which reflect the criticality of the pipeline system. The time to the first inspection and frequency of future inspections should be determined based upon various factors such as the potential corrosive effect of the fluid, the detection limit and accuracy of the inspection system, results from previous inspections, and changes in the operational parameters for the pipeline system. The following list may be used as a guide when determining the frequency and extent of these inspection surveys:

- inspection type;
- design and function of the pipeline system;
- seabed conditions;
- protection requirements;
- environmental conditions;
- corrosion and erosion conditions;
- third party traffic density;
- as-installed condition of the pipeline system;
- experience from previous inspection surveys;
- possible consequences of failure.

Periodical inspection – pipeline
The aim of the pipeline inspection survey is to verify that the design conditions are fulfilled, including:

- seabed stability (e.g. scour, sand waves);
- integrity of pipeline support (e.g. for free spanning pipeline sections);
- integrity of protection cover (e.g. mattresses, sand bags);
- integrity of weight coating;
- pipeline stability (e.g. excessive lateral pipe movements, upheaval buckling);
- pipeline expansion;
- extent of marine growth;
- security of mechanical connections, clamps and anodes;
- performance of external corrosion protection system;
- internal corrosion;
- integrity of valves, anchors, tee- and wye-connections.

A visual inspection of exposed pipeline sections is carried out to determine the general condition of the pipeline, and to locate areas that may require a more detailed inspection, including the detection and mapping of:

- mechanical damage to the pipe (NDT inspection may be required for the detection of cracks);
- damage to coating, insulation, possible jacket sleeves, field joints, etc.;
- anode consumption and condition, including integrity of electrical connections;
- seabed condition with respect to scouring or the build-up of seabed deposits;
- evidence of lateral and axial movements;
- buckled pipe sections (e.g. upheaval buckling);
- free pipeline spans;
- leaks;
- marine growth.

Concerning the last item, it is worth noting that the type and amount of biological activity may be used to assess the probable exposure time of otherwise buried lines.

The performance of the external corrosion protection system should be verified at regular intervals. Electric potential measurements of a cathodic protection system are normally carried out within one year of pipeline installation. The subsequent surveys may be carried out at greater intervals, provided the system is performing satisfactorily.

Measurements of the pipe wall thickness may be required where there is reason to believe that the pipe is subjected to severe corrosion or erosion.

Periodical inspection – riser

The riser should be inspected above water and below water to verify that the design conditions are fulfilled. The inspection survey should cover:

- exposure of buried or covered part of the riser base;
- scour at the riser base;
- riser displacements due to pipeline expansion;
- extent of marine growth;
- integrity of mechanical connections;
- integrity of protection cover (e.g. mattresses) at the riser base;
- coating damage;
- external and internal corrosion;
- performance of external corrosion protection system;
- integrity of riser supports.

The corrosion/erosion of the riser bore may be evaluated by carrying out wall thickness measurements at pre-selected reference points on a regular basis. The measurements are normally carried out using ultrasonic equipment. Special attention should be given to pipe bends and pipe reducers, which may be more susceptible to local corrosion and erosion attack than straight pipe sections. Furthermore, the riser bore may be subjected to inspection by lowering suitable equipment down inside the riser.

Chemical attack from bore fluids may be assessed using corrosion monitoring probes. When determining the location and position on the riser for the installation of these probes, attention should be given to pipe routing, product type and flow characteristics. The best accessible location to place the corrosion monitoring probes is on the topside part of the riser.

The performance of the external corrosion protection system should be verified at regular intervals for the riser. It is important that electrical potential readings are taken a sufficient distance away from adjacent anodes.

11.5   Repair assessment

11.5.1   *General*

When a defect or damage is reported, it is of utmost importance to maintain the operational condition (in particular pressure and temperature) at or below the condition at the time the defect or damage was found. As an example, after an upheaval buckling event, a constant fluid temperature and pressure should be maintained to avoid low-cycle, high-strain fatigue failure occurring at the apex of the buckle.

An assessment of the defect or damage should be made to determine the acceptability or requirement for pressure de-rating, repair or other corrective action. This includes consideration of whether the damaged pipeline system can be operated temporarily under the reduced operational condition until the defect has been removed. Also, one could consider installing a temporary repair clamp on the pipe until the permanent repair can be carried out.

The damage assessment should pay special attention to the risk of a sudden pipe failure. In particular, the possibility of personal injury, for example during diver inspections, may require the pressure to be reduced to a level which will not lead to pipe rupture. Following a repair, pressure testing should be performed for the repaired section. It is important that repairs be carried out by qualified personnel in accordance with approved specifications and procedures, and up to the standard defined for the pipeline.

11.5.2   *Pipe defects*

Grooves, gouges and notches
Sharp defects such as grooves, gouges and notches may be removed by grinding, or by other approved repair methods. It is important that the remaining pipe wall

thickness meets the minimum wall thickness requirement according to the pipeline standard. Deeper defects may be removed by cutting out the damaged part of the pipe as a cylinder, or alternatively, the pipeline may be re-qualified to a lower design pressure.

Dents

The criterion for accepting dents in a pipe should be evaluated in each case, considering:

- location, size and shape of the dent;
- imposed strains;
- properties of pipe material;
- composition of bore fluid;
- pressure and temperature;
- possible consequences of a pipe rupture;
- pigging possibility.

Special attention should be paid to dents affecting the welds in a pipe, as they may be critical for crack initiation/growth in the pipe wall. When a dent is found unacceptable it may be removed by cutting it out as a cylinder, followed by a welding repair. Alternatively, the damaged pipe section may be repaired by installing a full encirclement welded split sleeve or by the use of a mechanical coupling. If the dent cannot be accepted for the specified design pressure, the pipeline may be re-qualified to a lower design pressure.

Leaks

Leaks may be caused by insufficient contact pressure at the seal area in the flange and coupling connections. In this case, the seal integrity may be re-established by simply increasing the bolt pre-tension. This normally requires that the pipe system be depressurised so the operation can be carried out safely. However, prior to increasing the pre-tension it should be documented by calculations that no over-stressing will take place in the bolts, flange and gasket/seal. Alternatively, it may be required to remove the bolt pre-tension and clean the seal seats, and/or replace the gaskets and seals. Note that bolts that have been subjected to a high pre-tension may need to be replaced by new bolts.

If the leak is caused by cracks or corrosion in the pipe wall (weld), a suitable repair method needs to be established. The repair strategy depends on a number of factors: the pipe material, pipe dimension, location of leak, load conditions, pressure and temperature. The damaged pipe section may be repaired by installing a full encirclement welded split sleeve or by the use of a mechanical coupling. For some pipe applications (e.g. low pressure) it may be acceptable to install a repair clamp around the pipe, where the tightness is obtained by either welding, infill material, friction or by other mechanical means.

Bibliography and references

Introduction

The sections below list the documents referenced in the text. A distinction is made between Guidance Documents (codes, standards and recommendations) and Books and papers. The former are listed under the issuing organisations (in alphabetical order), whereas the latter are listed alphabetically under the (first) author.

Guidance documents

American Gas Association (AGA)
See Pipeline Research Council International, Inc. (PRCI).

American Petroleum Institute (API)
API 5 L *Specification for line pipe*
API 1104 *Welding of pipelines and related facilities*
API Recommended Practice IIII (1993) *Design, construction, operation and maintenance of offshore hydrocarbon pipelines*, 2nd edition.
API RP 17B (2002) *Recommended practice for flexible pipe*, 3rd edition.
API RP 17C *Recommended practice on TFL (Through FlowLine)*.
API Spec 17J (2002) *Specification for unbonded flexible pipes*.
API Spec 17K *Specification for bonded flexible pipe*.
API Technical Report TR17RUG (2002) *The ageing of PA-11 in flexible pipes*.

American Society of Mechanical Engineers (ASME)
ASME B31.3 (2002) *Process piping*.
ASME B31.8 (2003) *Gas transmission and distribution systems*.
ASME B31.8S (2001) *Managing system integrity of gas pipelines*.
ASME Section VIII (2001) *Boiler and pressure vessel code, pressure vessels, division 2*.

American Society for Testing and Materials (ASTM)
ASTM A262 *Standard practice for detecting the susceptibility to intergranular attack in austenitic stainless steels*.

ASTM C39/C 39M-01 *Standard test method for compressive strength of cylindrical concrete specimens.*
ASTM D422-63 *Standard test method for particle-size analysis of soils.*
ASTM G8 *Cathodic disbonding of pipeline coatings.*
ASTM G48 *Standard test methods for pitting and crevice corrosion resistance of stainless steels and related alloys by the use of ferric chloride solution.*
ASTM D4318-00 *Standard test method for liquid limit, plastic limit and plasticity index of soils.*

Asian Development Bank, Office of the Environment
Environmental guidelines for selected industrial and power development projects, 1990.

ASM international, The Materials Information Society
ASM Handbook No.1, (1990) 10th edition.

British Standards Institution (BS)
BS 4515 (2000) *Process of welding of steel pipelines on land and offshore.*
BS 7910 (1999) *Guidance on methods for assessing the acceptability of flaws in fusion welded structures.*
BS 8010 (1993) *Code of practice for pipelines. Part 3. Pipelines subsea: design, construction and installation.*
PD 8010-2 *Code of practice for pipelines. Part 2: Subsea pipelines* (replacing BS 8010).

Bureau Veritas
NI 364 DTO R00 E (1990) *Non-bonded flexible steel pipes used as flow-lines.*

Centre Européen de Normalisation (CEN)
EN 206-1 *Concrete – Part 1: Specification, performance, production and conformity.*
EN 1594 *Gas supply systems – Pipelines for maximum operating pressure over 16 bar – Functional requirements.*
EN 10204 *Metallic products – Types of inspection documents.*
EN 10208 *Steel pipes for pipelines for combustible fluids – Technical delivery conditions: Part 1. Pipes of requirement class A; Part 2. Pipes of requirement class B.*
EN 14161 *Petroleum and natural gas industries – Pipeline transportation systems (ISO 13623:2000 modified)* (in press).

Canadian Standardisation Association (CSA)
CSA Z662 – 94 Oil and gas pipeline systems.
   This standard supersedes these standards:

* CAN/CSA Z183 Oil pipeline systems
* CAN/CSA Z184 Gas pipeline systems
* CAN/CSA Z187 Offshore pipelines.

Dansk Standard (DS)
DS/R 464 (1988) *Corrosion protection of steel structures in marine environ-ments* (translation edition December 1993).

Det Norske Veritas (DNV)
DNV OS-F101 (2000) *Submarine pipeline systems.*
DNV RP B401 (1993) *Cathodic protection design.*
DNV RP C203 (2000) *Fatigue strength analysis of offshore steel structures.*
DNV RP E305 (1988) *On-bottom stability design of submarine pipelines,*
VERITEC.
DNV RP F105 (2002) *Free spanning pipelines.*
DNV RP F106 (2000) *Factory applied coatings for corrosion control.*
DNV RP F107 (2001) *Risk assessment of pipeline protection.*
DNV RP H101 (2003) *Risk management in marine and subsea operations.*
DNV (1987) *Guidelines for flexible pipes,* VERITEC.
DNV Guideline No 13 (1997) *Interference between trawl gear and pipelines.*
DNV Classification Notes No. 30.5 (2000) *Environmental conditions and envir-onmental loads.*
DNV (1994) *Rules for certification of flexible risers and pipes.*

Deutscher Institut für Normung (DIN)
DIN 2413 (1993) *Steel pipes, calculation of wall thickness subjected to internal pressure* (1972).
DIN 17 172 *Stahlrohre für Fernleitungen für brennbare Flüssigkeiten und Gase; Technische Lieferbedingungen.*
DIN V 19250/19251 *Control technology.*

European Federation of Corrosion (EFC)
EFC Publications Number 16 (1995) *Guidelines on materials requirements for carbon and low alloy steels for H_2S-containing environments in oil and gas production,* The Institute of Materials, London.
EFC Publications Number 17 (1997) *Corrosion resistant alloys for oil and gas production: guidelines on general requirements and test methods for H_2S ser-vice,* The Institute of Materials, London.

European Pipeline Research Group (EPRG)
Guidelines for defect acceptance levels in transmission pipeline girth welds.

European Union (EU)
Council Directive 97/11/EC of 3 March 1997 on the assessment of the effects of certain public and private projects on the environment, Official Journal No L 073, 14/03/1997 P 0005.

Gas Piping Technology Committee (GPTC)
ANSI/GPTC Z380.1-1998 *Guide for gas transmission and distribution piping systems,* 1998–2000.

Germanischer Lloyd (GL)

Rules for classification and construction, III – Offshore technology, Part 4 – Subsea pipelines and risers (1995).

Health and Safety Executive

PARLOC 94 (1996) *The update of loss of containment data for offshore pipelines* OTH 95 468, Advanced Mechanics & Engineering Ltd.

International Electrotechnical Commission

IEC 61508 *Functional safety of electrical/electronic/programmable electronic safety-related systems.*

International Standardisation Organisation (ISO)

ISO 3183-1 (1996) *Petroleum and natural gas industries – Steel pipe for pipelines – Technical delivery conditions – Part 1: Pipes of requirement class A.*

ISO 3183-2 (1996) *Petroleum and natural gas industries – Steel pipe for pipelines – Technical delivery conditions – Part 1: Pipes of requirement class B.*

ISO 3183-3 (1996) *Petroleum and natural gas industries – Steel pipe for pipelines – Technical delivery conditions – Part 1: Pipes of requirement class C.*

ISO 4287/1 *Surface roughness and its parameters.*

ISO 8501-1 *Preparation of steel substrates before application of paints and related products – Visual assessment of surface cleanliness – Part 1: Rust grades and preparation grades of uncoated steel substrates and of steel substrates after overall removal of previous coatings.*

ISO 9000 *Quality Management and Quality Assurance Standards.*

ISO 10474 *Steel and steel products – Inspection documents.*

ISO 13623 (2000) *Petroleum and natural gas industries – Pipeline transportation systems.*

ISO 14000 *Environmental management standard series.*

ISO 13847 (2000) *Petroleum and natural gas industries – Welding of pipelines.*

ISO 14723 (2001) *Petroleum and natural gas industries – Subsea pipeline valves.*

ISO 15156 *Petroleum, petrochemical and natural gas industries – Materials for use in H_2S-containing environments in oil and gas production, Parts 1–3.*

ISO 15741 *Paints and varnishes – Friction-reduction coatings for the interior of on- and offshore pipelines for non-corrosive gases.*

ISO 15589-2 *Petroleum and natural gas industries – Cathodic protection for pipeline transportation systems – Part 2: Offshore pipelines.*

National Association of Corrosion Engineers (NACE)

NACE MR 01-75 *Sulphide stress cracking resistant metallic materials for oilfield equipment.*

NACE RP 0492 (1992) *Metallurgical and inspection requirements for offshore pipeline bracelet anodes.*

NORSOK

M-503 (1997) *Cathodic protection*, Rev 2.

M-506 (1998) *CO_2 corrosion rate calculation model*, Rev 1.

U-102 *Remotely Operated Vehicle (ROV) services*, 2003-11-11.

Pipeline Research Council International, Inc. (PRCI)

Submarine pipeline on-bottom stability. Analysis and design guidelines (September 1993).

Pipeline Working Group of the Offshore Soil Investigation Forum (PWG OSIF)

Guidance notes on geotechnical investigations for marine pipelines, Rev 03 (1999).

United Nations Economic Commission for Europe (UNECE)

Convention on Environmental Impact Assessment in a Transboundary Context (1999).

US Military

Specification MIL-A-18001 (1987) *Anodes, corrosion preventive, zinc; slab, disc and rod shaped*, 1983 + Amendment 1.

World Bank, Environment Department

Technical Paper Number 14, (1991) *Environmental Assessment Sourcebook, Volume III, Guidelines for environmental assessment of energy and industry projects.*

Books and papers

Andersen, M., Berg, A. and Saevik, S. (2001) Development of an optical monitoring system for flexible risers, *Offshore Technology Conference*, Houston, Texas, OTC 13201.

Ayers, R.R., Allen, D.W., Lammert, W.F., Hale, J.R. and Jacobsen, V. (1989) Submarine pipeline on-bottom stability: recent AGA research, *Proceedings of Eighth Offshore Mechanics and Arctic Engineering Conference*, The Hague, OMAE, Volume 5, pp. 95–102.

Braestrup, M.W. (1989) Statical analysis of free spans in submarine pipelines, *Bygningsstatiske Meddelelser*, Vol. 60, No 2, pp. 51–70.

Braestrup, M.W. (1993) Shakedown stress analysis of pressurized pipelines subjected to bending, *Twelfth International Conference on Offshore Mechanics and Arctic Engineering*, Glasgow 20–24 June 1993, 10 pp.

Bryndum, M.B., Jacobsen, V. and Brand, L.P. (1983) Hydrodynamic forces from wave and current loads on marine pipelines, *Proceedings of the Fifteenth Annual Offshore Technology Conference*, Houston, Texas, OTC 4454, May 1983, pp. 95–102.

Bryndum, M.B., Jacobsen, V. and Tsahalis, D.T. (1992) Hydrodynamic forces on pipelines: model tests, *Journal of Offshore Mechanics and Arctic Engineering*, Vol. 114, No 4, pp. 231–241.

Christensen, C. and Hill, R.T. (1988) Hydrogen uptake considerations in cathodically protected structures in seawater, *Second NACE International Conference*, Milan, 4 pp.

Christensen, C. and Ludwigsen, P.B. The deleterious effect of cathodic protection on mechanically damaged pipes. In *Proceedings of International Conference on Offshore Mechanics and Arctic Engineering*, The Hague, 19–23 March 1989, Vol. V, pp. 377–388.

Dahl, W. Werkstoffkundlichen Grundlagen zum Verhalten von Schwefel im Stahl, *Stahl und Eisen*, Vol. 97, 1977, pp. 402–409.

de la Mare, R.F. (Ed.) (1985) *Advances in Offshore Oil and Gas Pipelines*, Oyez Scientific and Technical Services Ltd, London, 381 pp.

de Waard, C. and Milliams, D.E. (1975) Prediction of carbonic acid corrosion in natural gas pipelines, *First International Conference on the Internal and External Protection of Pipes*, Paper F1, University of Durham.

Denys, R. (1995) Pipeline technology. In *Proceedings of the Second International Pipeline Technology Conference*, Ostende, Belgium, September 11–14, Elsevier Science B.V., Amsterdam.

Driskill, N.G. (1981) *Review of expansion and thermal growth problems in subsea pipelines*, Offshore Oil and Gas Pipeline Technology, European Seminar, 3–4 February.

Grace, R.A. and Zee, G.T.Y. (1981) Wave forces on rigid pipes using ocean test data, *Journal of the Waterway, Port, Coastal and Ocean Division*, ASCE, **107**, WW2, 71–92.

Heller, W. (1966) Einfluss der chemischen Zusammensetzung auf die mechanischen Eigenschaften von unlegierten und niedriglegierten Stählen, *Stahl und Eisen*, **86**, 42–46.

Jacobsen, V., Bryndum, M.B. and Fredsøe, J. (1984) Determination of flow kinematics close to marine pipelines and their use in stability calculations, *Proceedings of the Sixteenth Annual Offshore Technology Conference*, Houston, Texas, OTC, Volume 3, pp. 481–492.

Jones, W.T. (1971) Forces on submarine pipelines from steady currents, ASME paper 71-UnT-3, presented at the Petroleum Mechanical Engineering with Underwater Technology Conference, Houston, Texas, September 1971.

Kastner, W., Roehrich, E., Schmitt, W. and Steinbuch, R. (1981) Critical crack sizes in ductile piping, *International Journal of Pressure Vessels and Piping*, **9**, Issue 3, pp. 197–219.

Krass, W., Kittel, A. and Uhde, A. (1979) *Pipelinetechnik*, Verlag TÜV Rheinland GmbH, Köln.

Lambrakos, K.F., Chao, J.C., Beckmann, H. and Brannon, H.R. (1987) Wake model of hydrodynamic forces on pipelines, *Ocean Engineering*, **14**, 2.

Loeken, P.A. (1980) The 'creep' on the Ekofisk – Emden 26″ Gas Pipeline, *Proceedings of the 12th Annual Offshore Technology Conference*, Houston, Texas, OTC 3783, May 1980, pp. 393–401.

Luo, D., Franklin, J. and Wright, A. (2001) Theoretical model of water ingress to pipe coating materials. In *Fourteenth International Conference on Pipeline Protection* (Ed. J. Duncan) BHR Group Ltd.

Miles, D.J. and Calladine, C.R. (1999) Lateral thermal buckling of pipelines on the sea bed, *Journal of Applied Mechanics*, **66**, 891–897.

Miller, A.G. (1988) Review of limit loads of structures containing defects, *International Journal of Pressure Vessels and Piping*, Vol. 32, Issues 1–4, pp. 197–327.

Mouselli, A.H. (1981) *Offshore Pipeline Design, Analysis, and Methods*, PennWell Publishing, Tulsa, Oklahoma, 193 pp.

Nielsen, N.J.R., Lyngberg, B. and Pedersen, P.T. (1990a) Upheaval buckling failures of insulated buried pipelines: a case story. In *Proceedings of the Offshore Technology Conference*, Houston, 7–10 May 1990.

Nielsen, N.J.R., Pedersen, P.T., Grundy, A. and Lyngberg, B.S. (1990b) Design criteria for upheaval creep of buried sub-sea pipelines, *Journal of Offshore Mechanics and Arctic Engineering (OMAE)*, **112**, November.

Nielsen, N.J.R., Pallesen, T.R. and Lyngberg, B. (1991) Installation inspection in relation to upheaval buckling, Offshore Pipeline Technology (OPT) European Seminar, Copenhagen, 14, 15 February 1991.

Palmer, A.C. and Ling, M.T.S. (1981) Movements of submarine pipelines close to platforms, *Proceedings of the 13th Annual Offshore Technology Conference*, Houston, Texas, OTC 4067, Vol. 3, May 1981, pp. 17–24.

Palmer, A.C. and King, R.A. (2004) *Subsea Pipeline Engineering*, PennWell, 570 pp.

Rinebolt, J.A. and Harris, W.J. (1951) Effect of alloying elements on notch toughness of pearlitic steels, *Transactions of the A.S.M.*, **43**, 1175–1214.

Sarpkaya, T. and Rajabi, F. (1980) Hydrodynamic drag on bottom-mounted smooth and rough cylinders in periodic flow. In *Proceedings of the Eleventh Annual Offshore Technology Conference*, Houston, Texas, OTC 3371, May 1980, pp. 219–226.

Sriskandarajah, T. and Mahendran, I.K. (1987) Parametric consideration of design and installation of deepwater pipelines, Offshore Oil and Gas Pipeline Technology, European Seminar, 35 pp.

Trautmann, C.H., O'Rourke, T.D. and Kulhawy, F.H. (1985) Uplift force–displacement response of buried pipe, *Journal of Geotechnical Engineering*, **III**, 9.

Verley, R., Lambrakos, K.F. and Reed, K. (1987) Prediction of hydrodynamic forces on seabed pipelines. In *Proceedings of the Nineteenth Annual Offshore Technology Conference*, Houston, Texas, OTC 5503, April 27–30 1987.

Werner, D.P., Coulson, K.E.W. and Barlo, T.J. (1991) Tests show barrier coatings do not block cathodic protection, *Oil and Gas Journal, Technology*, 14 October, 80–84 (Corrections, 11 Nov 1991, p. 65).

Wolfram, W.R., Getz, J.R. and Verley, R. (1987) PIPESTAB project: improved design basis for submarine pipeline stability. In *Proceedings of the Nineteenth*

Annual Offshore Technology Conference, Houston, Texas, OTC 5501, April 27–30 1987.

Yong Bai (2001) *Pipelines and Risers*, Elsevier Ocean Engineering Book Series, Volume 3, 498 pp.

Author index

The index below identifies all pages, including the Bibliography and references, where cited authors are mentioned.

Subject index

The layout of the book in chapters, sections and sub-sections is intended to facilitate the search for particular subjects. The alphabetic index below, however, also identifies locations that are not self-evident. Page numbers in **bold** indicate a principal treatment of the entry, whereas numbers in *italics* refer to figures. For explanation of terms see pages xxi–xlii of the Glossary and notation, which includes synonyms that are not listed below. Reference is also made to the listing of abbreviations (pages xvii–xxi), which identifies sections where further information can be found.

Printed and bound in the UK by
CPI Antony Rowe, Eastbourne